Introduction to Photocatalysis

From Basic Science to Applications

Introduction to Photocatalysis

From Basic Science to Applications

Yoshio Nosaka

Nagaoka University of Technology, Japan
Email: nosaka@nagaokaut.ac.jp

Atsuko Nosaka

Nagaoka University of Technology, Japan
Email: aynosaka@nagaokaut.ac.jp

THE QUEEN'S AWARDS
FOR ENTERPRISE:
INTERNATIONAL TRADE
2013

Print ISBN: 978-1-78262-320-5

A catalogue record for this book is available from the British Library

Published by The Royal Society of Chemistry,
Thomas Graham House, Science Park, Milton Road,
Cambridge CB4 0WF, UK

Registered Charity Number 207890

Visit our website at www.rsc.org/books

Printed in the United Kingdom by CPI Group (UK) Ltd, Croydon, CR0 4YY, UK

Preface

The photocatalysts can decompose stains with the aid of photo energy and have been recognized as being useful for environmental clean-up. They have become rapidly widespread worldwide. The technique was originally developed in Japan with the use of the semiconductor of titanium dioxide aiming at the production of hydrogen by water decomposition. Numerous researches on this subject have begun so far. Several thousand scientific reports on photocatalysts are currently published every year, and it would be almost impossible to look over all of them. However, the proper selection of the necessary issues from this enormous amount of literature would be relatively feasible owing to the development of information systems. However, problems arise over whether the authors correctly refer to the past researches when interesting and relevant reports are found. For instance, almost the same research as Fujishima *et al.* reported in 1975 was published in a prestigious journal, *Science*, 25 years later. In the meantime, four comments that pointed out the mistake in judging the acceptance of the reports were published in *Science* as well (A. Fujishima, *Science*, 2003, **301**, 1673). Among the literature in the field of photocatalysts, a lot of reports whose results were not reproducible were published in the prestigious journals. One of the reasons why such incorrect papers appear is that the science which it is based on includes many varied fields from physics to chemistry or biology. Photocatalysts are somewhat different from general catalysts, especially because it is currently recognized that the photocatalytic reactions have been developed as the extension of electrode reactions. Therefore, although they are a kind of catalyst, the standpoints are often different depending on the research bases. The researches

Introduction to Photocatalysis: From Basic Science to Applications
By Yoshio Nosaka and Atsuko Nosaka
© Yoshio Nosaka and Atsuko Nosaka, 2016
Published by the Royal Society of Chemistry, www.rsc.org

related to photocatalysts cover the research fields such as photo-chemistry, electrochemistry, catalyst chemistry, radiation chemistry, physical chemistry, organic reaction chemistry, solid-state physics, film science, ceramics science, materials science, *etc.* This book deals with photocatalysis, where generally speaking the photocatalysis is the heterogeneous photocatalysis using semiconductors. As for homogeneous photocatalysis, it belongs to the field of traditional photochemistry, and distinguished textbooks explaining from this basis are available.

Although enormous amounts of research papers on photocatalysts have been published, few introductory textbooks from the scientific viewpoints are presented so far because the field of photocatalysis covers various fields of science as stated above. Therefore, an intro-ductory textbook aiming at understanding the foundation of photo-catalysis is indispensable. To introduce scientific bases of photocatalysis to graduate students, no proper textbook seems available, although many specialized books have been published. We have published a book named "*Introduction to Photocatalysis*" written in Japanese from Tokyo Tosho, Inc. in 2004. Because a lot of new reports were later published, we felt the necessity of a drastic revision of the book. Meanwhile, the RSC asked us to write a textbook on photocatalysis. Thus, we were willing to publish this book. The object of this book is providing the students and researchers from the faculty of engin-eering not with novel reviews but mainly with the science. This sci-ence will become the base of photocatalysis and provide the way of thinking that makes it easy to understand the concepts towards the practical application of photocatalysts as functional materials.

The number of reports on photocatalysis is still growing. Thus, targeting the subjects in a diversified field, we referred to many books and specialized journals. Our concern is that some reports contra-dictory to those referred to in this book might appear, which might cause misunderstanding. We would appreciate it if readers could respond positively and give us their honest opinions.

We are grateful to Mr. Masahide Mase of Gijutsu-kyouiku-shuppan (Photo Functionalized Materials Society, PFMS) for promoting to publish the previous Japanese textbook on photocatalysis, and Mr. Tomoyoshi Matsunaga of Tokyo Tosho for approving to use the figures and the illustrations.

Yoshio Nosaka & Atsuko Nosaka
Nagaoka University of Technology

Symbols and Acronyms

a	radius of a particle
A	relative absorption, or surface area
A	electron accepter
ACS	American Chemical Society
AM	air mass
AQY	apparent quantum yield
α	absorption coefficient
b	interatomic distance
c	speed of light in vacuum $(=3.00 \times 10^8\ \text{ms}^{-1})$
CB	conduction band
COD	chemical oxygen demand
d	film thickness
D	electron donor, or dye
DOS	density of states
e	elementary charge $(=1.602 \times 10^{-19}\ \text{C})$
e^-	photoinduced electron in semiconductor
ε^*	complex permittivity $(\varepsilon^* = \varepsilon_1 + i\varepsilon_2)$
ε_0	permittivity of vacuum $(=1/c^2\mu_0)$
ε_1	real part of complex permittivity (dielectric constant)
ε_2	imaginary part of complex permittivity
E	electric field, or energy
E^0	standard redox potential
E_g	bandgap energy
E_F	Fermi energy level
E_C	energy level of conduction band edge
E_V	energy level of valence band edge

Introduction to Photocatalysis: From Basic Science to Applications
By Yoshio Nosaka and Atsuko Nosaka
© Yoshio Nosaka and Atsuko Nosaka, 2016
Published by the Royal Society of Chemistry, www.rsc.org

ESR	electron spin resonance (=electron paramagnetic resonance, EPR)
FTIR	Fourier transform infrared spectroscopy
ϕ	electric potential, or quantum yield
h	Planck constant ($=6.63 \times 10^{-34}$ Js) ($\hbar = h/2\pi$), or thickness of film
h^+	photoinduced hole in semiconductor
HRTEM	high-resolution transmission electron microscopy
η	viscosity
I	intensity (power) of light in unit of mW cm^{-2} (1 mW cm^{-2} = 10 W m^{-2})
IR	infrared
IPCE	incident photon-to-current efficiency
k	reaction rate constant, imaginary part of complex refractive index (=absorption index), wave vector of Bloch function
k_{B}	Boltzmann constant ($=1.381 \times 10^{-23}$ JK^{-1})
K	equilibrium constant
L_{sc}	width of space charge layer
LED	light emitting diode
LIF	laser induced fluorescence
λ	wavelength of light
m	specific refractive index ($=n^*/n_{\mathrm{m}}$)
m^*	effective mass
M	unit of molar concentration (mol dm^{-3}), or metal
MB	methylene blue
μ	magnetic permeability of substance, or mobility of carrier
μ_0	magnetic permeability of vacuum ($=4\pi \times 10^{-7}$ Hm^{-1})
μ^*	reduced effective mass
n	real part of complex refractive index
n^*	complex refractive index ($n^* = n + \mathrm{i}k$)
n_{m}	refractive index of medium
N_{A}	Avogadro number ($=6.02 \times 10^{23}$ mol^{-1})
N_0	carrier density, or density of photos absorbed in semiconductor
NMR	nuclear magnetic resonance
ν	frequency of electromagnetic wave
ppmv	parts per million by volume
r	reaction rate
R	relative reflectance, or rejection ratio
R_{A}	reduction rate of electron acceptor A
R_{D}	oxidation rate of electron donor D

RHE	reversible hydrogen electrode
RSC	Royal Society of Chemistry
ρ	charge density, or density of materials
SC	semiconductor
SHE	standard hydrogen electrode (=NHE, normal hydrogen electrode)
T	relative transmittance, or absolute temperature
TOC	total organic carbon
TPD	temperature programmed desorption
θ	contact angle
u	speed of substrate
U_{fb}	flat band potential
US	ultrasonic wave
UV	ultraviolet (light)
υ	velocity of light in materials
V	kinetic energy
VB	valence band
Vis	visible light
VOC	volatile organic compound
ω	angular frequency (=$2\pi\nu$), or angular speed of rotation
χ	particle size relative to the wavelength of light ($\chi = 2\pi a/\lambda$)
XAFS	X-ray absorption fine structure
XPS	X-ray photoelectron spectroscopy
XRD	X-ray diffractometry

Contents

Introduction to Photocatalysis: From Basic Science to Applications
By Yoshio Nosaka and Atsuko Nosaka
© Yoshio Nosaka and Atsuko Nosaka, 2016
Published by the Royal Society of Chemistry, www.rsc.org

3 Principles of Semiconductors   45

4 Principles of Photoelectrochemistry   65

5 Photocatalyst Surface and Active Species   84

6 Kinetics and Mechanism in Photocatalysis   111

10   Future Applications of Photocatalysis   232

1 What is a Photocatalyst?

1.1 Function and Utilization of Materials

Materials supporting our modern lives are composed of various combinations of numerous substances. The substances are mostly elements or chemical compounds, which are often in solid states. Different from liquids and gases like water and air, solids do not always require any container to shape them, because they can make a shape by themselves. Therefore, we often presume that most substances around us are solids. Generally speaking, the following three fundamental conditions of materials and substances must be fulfilled for practical use.

1. It possesses characteristic properties or functions that cannot be found for the other materials.
2. It is harmless for human beings and natural environments.
3. The raw materials must be abundantly and readily obtainable and they can be produced with reasonable price.

Now, let's look at how TiO_2 fulfills these conditions as a useful and practical material in our daily lives.

Every material consists of various atoms which are categorized by elements. The number of elements on Earth is about 100. The major material of usual chemical substances mostly consists of from one to three elements. Taking into account the number of the combination of these elements, a variety of solids would be formulated. Now the question is what are the states of the atomic nuclei and electrons in the solid when we remember that an atom consists of a positively

Introduction to Photocatalysis: From Basic Science to Applications
By Yoshio Nosaka and Atsuko Nosaka
© Yoshio Nosaka and Atsuko Nosaka, 2016
Published by the Royal Society of Chemistry, www.rsc.org

charged nucleus and several negatively charged electrons surrounding the nucleus. Because the atomic mass is mostly determined
by the weight of the nuclei, the state of substance changes when the
nucleus moves. On the other hand, electrons are very light and play a
role to connect distinct nuclei. But they cannot always move around
within the solid. The arrangement of the atoms determines the
character of the solid. In a crystalline state, the atoms are arranged
regularly. Otherwise, the state is called non-crystalline (amorphous).
In the crystalline form, the atomic defect disappears and the distance
of electron movement becomes prolonged. On the other hand,
amorphous materials like glass are randomly directed and are then
expected to easily form a smooth film. According to the intrinsic
characteristics and formulation, the individual functions of the
materials emerge to be utilized as substances which possess various
mechanical, electric, magnetic, and optical functions. To improve
these characteristics, novel materials have been designed and
developed.

1.2 Titanium Dioxide TiO$_2$ – a Representative Photocatalyst

A representative photocatalyst, titanium dioxide (TiO$_2$) is a white inorganic compound, which is stable and non-toxic. It is characterized
by the extremely large refractive index, dielectric constant, and insu-
lation resistance. TiO$_2$ has been utilized widely as a pigment because
the natural abundance of Ti is as high as ninth among the existing
elements on Earth and the price is relatively low. It is used for de-
lustering white cloth such as shirts, fillers for papers, plastics, paints,
and white rubber. However, few people know that an average person
usually wears several grams of titanium oxide.

The recent trends of the applications of TiO$_2$ are shown in
Figure 1.1. Nowadays, 89% of TiO$_2$ are used as pigments for many
products by taking advantage of the considerably high refractive
index. In addition, it is used for capacitors as electric ceramics by
exploiting the high dielectric constant. Recently, as a raw material of
glass and barium titanate, it has been utilized for capacitors and
various sensors.

The annual global consumption of titanium dioxide is 4 million
tons. The amount of the production increases with GNP. It has been
recently demonstrated to be useful for our daily life because of
the photocatalytic functions caused by the chemical reactions by

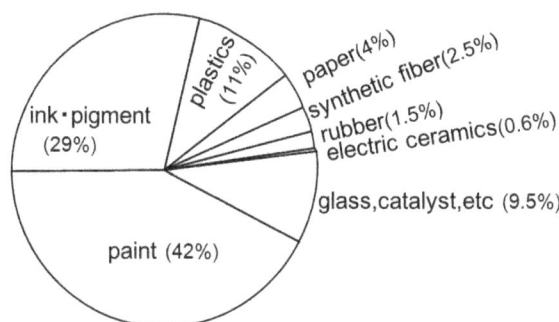

Figure 1.1 Proportion of the use of TiO₂.
(Japan Titanium Dioxide Industrial Association, Sept. 2010).

absorbing UV light contained in sunlight or a fluorescent lamp. The functions have opened up wider applications of TiO_2 in various fields, such as environmental clean-up, phototherapy, antifogging, and self-cleaning of mirrors and glasses. Now, let's look at why TiO_2 came to be used as a material in the wide field.

We will see the points of TiO_2 superior to the other materials. The representative properties of TiO_2 and ZnO photocatalysts, such as density (specific gravity), specific heat (thermal capacity), hardness, melting point, refractive index, dielectric constant, bandgap, mobility of the semiconductor carrier, and isoelectric point are shown in Table 1.1. There are two kinds of common crystalline system (rutile and anatase), which differ in the arrangements of atoms and show some different properties. The thermal capacity and density of TiO_2 are not particularly different from those of the other materials. However, since the hardness of rutile crystalline is remarkably high, TiO_2 is categorized as a hard material. Furthermore, the high melting point and heat resistance are advantages as materials. A more important advantage would be the high refractive index and the relatively wide bandgap, which is the reason why TiO_2 has been widely used in daily materials. Besides, although not presented in the table, it is also a prominent characteristic of TiO_2 that it is chemically so stable that it cannot be dissolved even in concentrated sulfuric acid, except on heating.

As shown in Table 1.1, the refractive index of each crystalline form is different. The average value is 2.72 and 2.52 for rutile and anatase, respectively. These values are even larger than that for diamond (2.42), and the largest among the minerals which are transparent for visible light. This means that the incident light is refracted largely at the surface of the solid. As will be described in Chapter 2,

Table 1.1   Properties of TiO$_2$ and ZnO crystals.[1–3]

Properties at 298 K	TiO$_2$ (Rutile)	TiO$_2$ (Anatase)	ZnO (Wurtzite)
Density (g cm^{-3})	4.250	3.894	5.606
Volume (nm^3 molecule^{-1})	0.0312	0.0341	0.0241
Specific heat (J K mol^{-1})	55.06	55.52	43.9
Mohs hardness	7.0–7.5	5.5–6.0	4.5
Melting point (°C)	1840 (decomp.)	trans. to rutile	1970 (decomp.)
Refractive index E//c	2.616	2.554	2.02
(@n$_D$, 589 nm) E$\perp$c	2.903	2.493	2.02
Relative permittivity, $\varepsilon(0)$	167 (//c) 86 ($\perp$c)	12	8.5 (//c) 7.4 ($\perp$c)
Bandgap energy (eV)	3.0 (direct)	3.8 (direct) 3.2 (indirect)	3.4
Effective mass (hole)	20	0.8	0.24
Mobility (cm^2 Vs^{-1})	0.1	4–20	130–205
Isoelectric point	5.6	6.1	10.3

the reflection (scattering) of the light on the surface of the solid theoretically becomes large with increasing refractive index. The scattering of the light depends on the particle size. As will be shown in Chapter 2, the scattering coefficient of the visible light becomes the largest when the particle size of TiO$_2$ reaches 250 nm. The relatively wide bandgap means that TiO$_2$ does not absorb visible light. Namely, because the light reflected diffusely is colorless, it looks white. By exploiting these characteristics, TiO$_2$ is used as a pigment to make the substance opaque.

For example, on coloring transparent plastics, the plastics are made opaque with TiO$_2$ and then colored. Papers used for dictionaries are extremely thin and almost transparent so that the characters on the back side appear on the front side. To make them opaque, a large amount of TiO$_2$ is added in the process of making paper pulp. It is also used to make opaque a roll paper of cigarettes, white rubber, and printing ink. The natural fibers scatter the light and produce the natural texture because of the unevenness of the surface. On the other hand, the surface of the chemical fibers is very smooth, because it reflects the light to present an unnatural gloss. Besides, the light scattering by the cloth diminishes in the wet state and it becomes transparent because the refractive index of chemical fibers is close to that of water. This causes some problems. Then, TiO$_2$ particles were incorporated into the production procedure of synthetic fibers. By the incorporation of TiO_2, the fibers start to scatter light and can be made opaque, and then provide a natural texture like natural fibers. Furthermore, when it is used as a pigment not only for white paints but also colored paints, the prominent diffusive light reflection leads to better coloring.

Every semiconductor that absorbs light can be used in principle as a photocatalyst. However, actually only TiO_2 is used practically at present. Although CdS and ZnO are also photocatalytic semiconductors, Cd and Zn are dissolved by oxidation in the absence of a compound which can be easily oxidized by the photocatalytic reactions. Therefore, if we used such photocatalysts, they would cause heavy metal pollution. TiO_2 is chemically so stable that it does not cause such dissolution. There are some other compounds that do not cause such dissolution. But the bandgaps are larger than that of TiO_2, which requires ultraviolet (UV) light with higher energy for photocatalytic reactions. TiO_2 does not absorb visible light but absorbs UV light close to visible light. Therefore, it is practically advantageous that the small amount of UV light contained in sunlight or fluorescent light can be used for the photocatalysis by TiO_2.

Besides, as described in the Japanese Pharmacopoeia, the safety of TiO_2 to humans is confirmed and it is permitted as a food additive. When used for environmental treatments, it is practically a prerequisite that the materials used are not harmful for the environment but have a strong ability for the treatments. Since the environmental treatment does not produce the added value, expensive materials cannot be employed. The natural abundance of titanium is rather high among the elements on Earth and the cost is reasonably low. TiO_2 for a photocatalyst costs US$20–60 kg^{-1} (for pigments, several dollars kg^{-1}). TiO_2 is the only material that fulfills these requirements and can be used as a photocatalyst at present. This is the reason why TiO_2 is extensively used as a photocatalyst.

1.3   Crystal Structure and Characteristics of TiO_2

There are three common crystal structures for TiO_2, *i.e.*, rutile, anatase, and brookite. The ore of rutile is usually obtained as a reddish brown crystal. The name is originated from *rutil*, which means "red" in Greek. Pure TiO_2 of any crystalline form is colorless. Rutile presents a red color because it contains impurities such as iron oxide. The origin of the word anatase is "upward tension" in Greek because the crystalline form is long. Brookite was named after an English mineralogist, H. J. Brook. Thermodynamically, rutile is in the stable phase within the whole temperature region and the other two are in the metastable phase. This means that, at higher temperatures, all TiO_2 is transformed to the rutile crystal. Brookite is more unstable than the other crystals and difficult to fabricate as a pure crystal form. Rutile is widely used for a pigment while anatase is mainly used for photocatalysts.

The crystal structures of metal oxides are analyzed usually by means of XRD (X-ray Diffraction). Because the wavelength of X-rays is close to the interatomic distance Ti–O (about 195 pm), the crystal structure is analyzed based on the principle that the reflection angle of the irradiated X-ray is different depending on the arrangements of the atoms. On the XRD chart, in most cases the intensities of the X-ray diffraction are plotted as a function of 2θ, where 2θ can be obtained from the spacing of the atom plane, d, and wavelength, λ, of an X-ray using the Bragg condition of $2\theta = 2\sin^{-1}(\lambda/2d)$. 2θ values of main peaks observed for two common polymorphs of TiO_2 are shown in Table 1.2, together with those of ZnO as a reference.

The stable ionic states of Ti and oxygen O are Ti^{4+} and O^{2-}, and the ionic radii are 74 and 122 pm, respectively. The ratio of the radii is 0.59. Because it is between 0.414 and 0.732, TiO_2 takes 6-coordinate structure in which O^{2-} is close-packed on the six apices of a regular octahedron and Ti^{4+} is incorporated at the center. Like most of the other metal oxides, the crystalline structure is represented with a regular octahedron (TiO_6) as a basic unit. As shown in Figure 1.2, the

Table 1.2   Crystalline systems and their XRD data of TiO_2 and ZnO^a.

TiO_2 rutile[4]			TiO_2 anatase[5]			ZnO wurzite[6]		
Tetragonal			Tetragonal			Hexagonal		
$a=b=459$ pm			$a=b=379$ pm			$a=b=325$ pm		
$c=296$ pm			$c=951$ pm			$c=521$ pm		
$2\theta/°$	Intens.	Index	$2\theta/°$	Intens.	Index	$2\theta/°$	Intens.	Index
27.42	100	110	25.28	100	101	36.23	100	101
36.09	50	101	37.80	20	004	31.77	57	100
41.23	25	111	48.05	35	200	34.43	44	002
54.32	60	211	53.89	20	105	56.60	32	110
56.64	20	220	55.06	20	211	62.86	29	103
69.01	20	301	62.69	14	204	47.54	23	102

$^a 2\theta$: calculated for X-ray (CuKα, $\lambda = 154.056$ pm) commonly used in X-ray diffraction analysis. pm $= 10^{-12}$ m.

Rutile

Anatase

Wurtzite

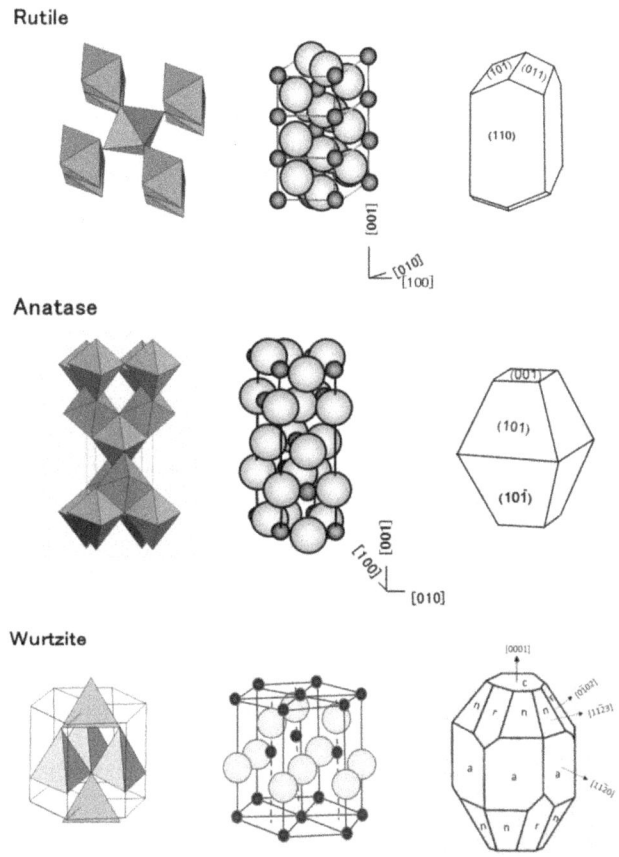

Figure 1.2 Crystal structures of rutile and anatase for TiO_2 and wurtzite for ZnO. Three unit cells are stacked for rutile. Small and large circles represent metal and O ions, respectively.

rutile structure is formed by sharing the two facing equatorial edges and also sharing the axial and horizontal oxygen atoms. On the other hand, when the regular octahedrons are packed without opening, the crystalline structure of sodium chloride is obtained. However, for this structure, the number of Ti atoms should be the same as that of O atoms. Then, when one Ti of every two positions in the sodium chloride structure is removed, the number of Ti becomes half of that of O, reflecting the actual number for TiO_2. Such an atomic arrangement is called anatase crystal structure. For a rutile crystal the two edges are co-owned, which extend towards the *c*-axis [001] successively and the chain structure is formed.

On the other hand, ZnO belongs to the hexagonal crystalline system with a wurtzite structure. Since the stable ionic state of Zn is Zn^{2+}, it takes 4-coordinate structure in which O^{2-} is close-packed on the four apices of a regular tetrahedron and Zn^{2+} is incorporated at the center. As shown in Figure 1.2, the crystal unit of rutile and anatase is a rectangular parallelepiped. Since the lengths of the *a*- and *b*-axes ([100] and [010] direction) are equal, they are called a tetragonal system. On the other hand, in wurtzite the angle between the *a*- and *b*-axes is 60° and it takes near hexagonal closest packed structure.

The specific volume of anatase is larger than that of rutile as shown in Table 1.1. Although the relative ratio of the ionic and covalent bond properties of the Ti–O bond is reported to be almost the same, the ionic properties diminish when the number of shared edges increases. Consequently, the ionic properties of rutile, in which the number of shared edges is two, are higher. For anatase, in which the number of shared edge is four, the ionic properties decrease. These crystal forms are considered to be closely related to the photocatalytic properties.

In general, the photocatalytic activity for anatase is higher than that for rutile. One of the reasons would be the difference of their bandgap structures. The bandgap of anatase is 3.2 eV and the energy of the conduction band edge is 0.2 eV higher than that of rutile (bandgap is 3.0 eV).[7] The photocatalytic reduction ascribable to this difference, especially the reduction of oxygen, takes place more readily for rutile than for anatase. One should take into account that the activities of rutile and anatase are different depending on the distinct reactions. The difference in the activities arising from the crystal structures will be described in detail in Chapter 7.

The surface of rutile TiO_2 crystal powder consists of mostly (110), (101), and (100) facets, and the (110) facet is reportedly 60% of the whole facets. In the previous reports, the surface of the anatase crystal particles was mostly comprised of (001) facets. However, in recent reports it is believed to be comprised of mainly (100), and (011) = (101) facets. For anatase crystals, the structure of the (011) and (101) facets is the same. Although small, the facets (010), (110), (111), (112), (113) are observed.[8] A large anatase crystal of decahedron structure with (101) and (001) facet was reported as shown in Figure 1.2. For comparison, the crystal structure of wurtzite ZnO is also shown in Figure 1.2, where a tetrahedron with centering Zn^{2+} can be the unit of the crystal.

The surface structure of anatase TiO_2 is shown in Figure 1.3A. The surface Ti^{4+} which is not fully coordinated is represented by a small closed circle (•). Two kinds of Ti^{4+} which are not fully coordinated were confirmed by investigating the adsorption of carbon monoxide

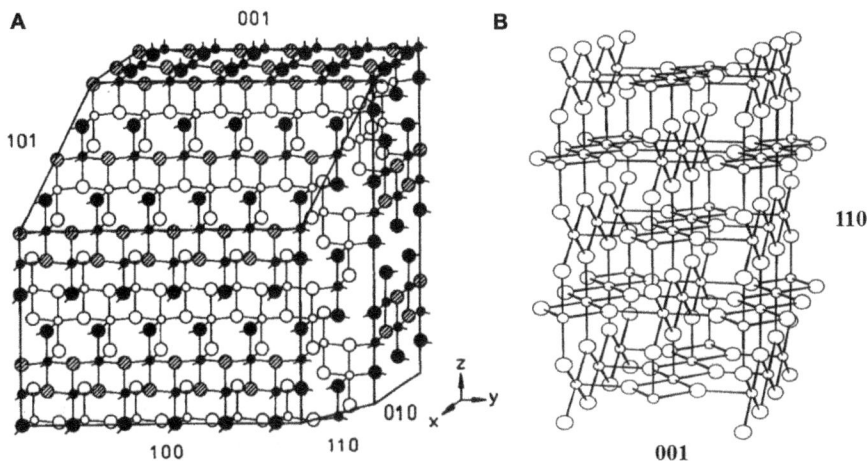

Figure 1.3 (A) Differently exposed surfaces of an anatase crystal.[8] Small circles ●○ represent Ti^{4+} and large circles ○● represent O^{2-}, respectively. Closed circles represent ions which locate on the surface and are not fully coordinated, while open circles represent those which locate on the subsurface. Hatched circles represent saturated O^{2-} which locate on the surface. Reproduced from ref. 8 with permission from the Royal Society of Chemistry. (B) Structure of rutile crystal.[9]
Reprinted from G. Pacchioni, A. M. Ferrari, and P. S. Bagus, Cluster and band structure *ab initio* calculations on the adsorption of CO on acid sites of the TiO_2 (110) surface, *Surf. Sci.*, **350**, 159–175. Copyright (1996), with permission from Elsevier.

(CO) by FTIR. These two Ti^{4+} can be distinguished by the difference of their electrophilicity. Namely, there are more attractive part α and less attractive part β. In Figure 1.3A, four coordinate Ti^{4+} of (110), (111), and (113) are α and five coordinate Ti^{4+} of (011)=(101) and (100)=(010) are β.[8]

1.4 Manufacturing of TiO_2

TiO_2 powder is industrially produced by two methods, that is, a sulfate process and a chlorine process. Raw materials utilized for titanium element are ores named ilmenite, rutile, and anatase. For the sulfate method, ilmenite ore is mainly used, while for the chlorine process, rutile ore, which is fabricated from ilmenite ore, or titanium slugs are used. The chlorine process, which causes less pollution problems because of its closed system, tends to be globally employed increasingly. For photocatalysts, TiO_2 is also manufactured industrially either by the sulfate or the chlorine processes. Besides, depending on the shapes of the material used as photocatalysts, TiO_2 is fabricated by various methods such as a sol–gel method which uses titanium alkoxide, and a spattering method by which very thin films can be obtained. The fabrication processes of TiO_2 specified for photocatalytic materials will be described in Chapter 7. In this section, the sulfate and chlorine processes which are employed industrially for the manufacturing of TiO_2 powder will be briefly introduced.

1.4.1 Sulfate Process

In Figure 1.4A, the manufacturing procedure of TiO_2 by a sulfate process is shown. In the sulfate process, ilmenite ore ($FeTiO_3$) is dissolved in the sulfuric acid.

$$FeTiO_3 + 2H_2SO_4 \rightarrow TiOSO_4 + FeSO_4 + 2H_2O \tag{1.1}$$

Then, $FeSO_4$ is removed followed by the condensation of dissolved $TiOSO_4$ solution. Afterwards, the solution is hydrolyzed with the addition of seed crystal. This procedure determines the particle size.

$$TiOSO_4 + mH_2O \rightarrow TiO_2\, nH_2O + H_2SO_4 \tag{1.2}$$

$$TiO_2\, nH_2O + (\text{calcination}) \rightarrow TiO_2 + nH_2O \tag{1.3}$$

TiO_2 manufactured with sulfuric acid is usually anatase crystal. Because of a small trace of the remaining sulfuric acid it does not transfer to the rutile phase even after calcination at $900\,°C$. To fabricate rutile crystal, rutile seeds are added in the hydrolysis procedure.[1]

Under specific hydrolysis conditions, hydrated titanium oxide $TiO_2 nH_2O$ can be obtained. Hydrated $TiO_2 nH_2O$ consists of amorphous ultrafine particles with large surface area. The detailed structure has not yet been elucidated but is suspected to be composed of very fine anatase or rutile crystallites. The hydrated titanium oxide has a large surface area but contains crystal defects which exert as recombination centers of electrons and holes. Therefore, it provides barely photocatalytic activity. Because the titanium oxide obtained by calcinating the hydrated titanium oxide contains impurities such as iron originated from raw materials and remaining sulfuric ions, it often shows low photocatalytic activity.[10]

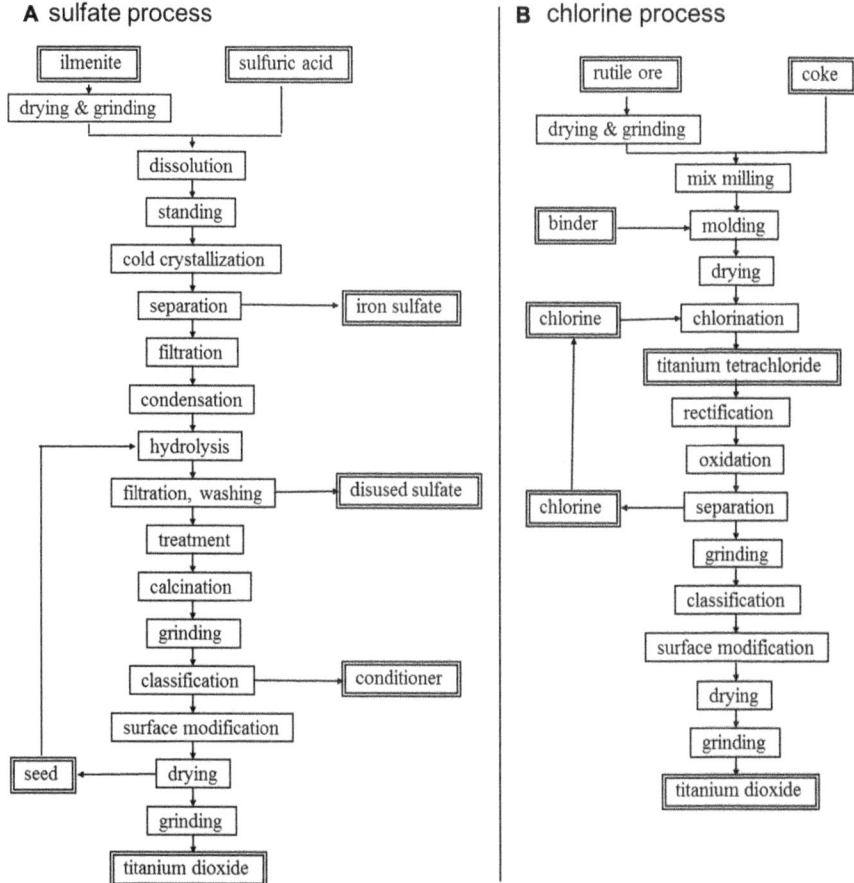

Figure 1.4 (A) Sulfate and (B) chlorine processes for manufacturing tita-
nium oxide.
Reproduced from *Sankachitan Hikarishokubai no Subete*
(K. Hashimoto and A. Fujishima, 1998) with permission from
CMC Publishing.

1.4.2 Chlorine Process

The chlorine process, which is shown in Figure 1.4B, is employed
preferentially for the fabrication of rutile crystal powder. At first, coke
and chlorine are added to rutile ore (FeO, TiO_2) to chlorinate.

$$TiO_2 + 2Cl_2 + C \rightarrow TiCl_4 + CO_2 \qquad (1.4)$$

After the removal of the generated iron chlorides ($FeCl_2$ and $FeCl_3$),
titanium tetrachloride ($TiCl_4$) is oxidized.

$$TiCl_4 + O_2 \rightarrow TiO_2 + 2Cl_2 \qquad (1.5)$$

The chlorine gas which is generated in this process can be used for the chlorination repeatedly. This is the eminent characteristic of the chlorine process. In addition, this method has an advantage in removing the impurities to obtain white powders more easily as compared to the sulfate process. Furthermore, because the procedure in which the particle formation takes place is performed in the gas phase, dispersive particles can be easily obtained. The characteristics of the refined TiO_2 reflect the component of the raw materials and the fabrication procedures. The surface of the titanium oxide fabricated by these methods is treated with Al_2O_3, SiO_2, or ZrO_2 depending on the applications.

Degussa P25 (Evonik or Japan Aerosil) which is often used in many literatures as a representative TiO_2 photocatalyst is fabricated by a chlorine process. It is a mixture of anatase and rutile and shows high photocatalytic activity in various reaction systems, which could be attributed to the reduced crystal defects and the enlarged surface area (about 50 $m^2 g^{-1}$) resulting from the treatment at high temperature.

The gas phase method is barely employed for fabrication in the laboratory because of the difficulties in the temperature control of the reaction and collection of the products. However, titanium tetrachloride, which is the intermediate of this reaction system, is used to prepare hydrated TiO_2 nanoparticles, or colloids of TiO_2 by hydrolyzation in the laboratory.

1.5 The History of Inactivation and Activation of TiO_2

Titanium oxide appeared for the first time in the report by W. Gregor, an English clergyman and an amateur chemist, who was asked to analyze a black stone (it is now named ilmenite) brought from Manaccan ravine in the county of Cornwall to obtain a novel metal oxide. Four years later, in 1795, a German chemist M. H. Klaproth discovered that rutile ore is an oxide of a new element. He named the element titanium after the giants of Greek mythology.[11] Titanium is contained in meteorites and the sun and it was reported that 12.1% TiO_2 was contained in the stone which was brought from the moon by the spaceship Apollo 17.[12]

As for the history of TiO_2 as a pigment, it was reported in 1913 that the coloring of various metal oxides on porcelain fabrication was improved by the use of TiO_2 as a pigment.[13] The production of the pigments of the mixture of the titanium oxides was initiated in Norway and the United States in 1916. It was gradually recognized that as white pigment TiO_2 is superior to the lead carbonate which had been used conventionally. In 1923, the production of pure anatase titanium oxide was initiated in France.[1] In the 1930s, the excellent whiteness was further recognized and many manufacturers were founded. In the early 1950s, the chlorine process was commercialized for the industrial production of rutile titanium oxide.[11] In the 1970s, the pollution problems attracted public attention and some factories which used the sulfate process were closed. Nevertheless, the production of the pigments was increasing. In the 1980s, it came to be produced for use other than white pigments. Because titanium oxide nanoparticles and hydrated titanium oxides possess properties such as absorption of UV light, transparency, and chemical stability, nowadays it is extensively used as indispensable materials in various fields such as foods, medicines, and cosmetics, although the amount is small.[1]

Though titanium oxide is mainly used as pigments in paints as stated above, when paint containing pure titanium oxide is applied and the exposure test is performed, chalking phenomena often take place. The chalking originates from the phenomenon that the paint forms powders like chalk to deteriorate under the sunlight and air. However, it was found in 1922 that very durable paint could be fabricated by mixing titanium oxides with ZnO.[14] Thus, titanium oxide was recognized a long time ago as a white pigment superior to the white lead (lead carbonate) but caused various undesired reactions under the light. The calking phenomena appear more often for anatase than for rutile. As a mechanism for chalking, it was reported already in the middle of 1970 that the active species such as superoxide, OH radicals, and singlet oxygen, which we consider now to be related to the photocatalytic reactions, were involved in the reactions.[15,16]

Titanium oxide has been manufactured for almost 100 years. Even now paints of various colors are produced by the combination of colored pigments with plastic resins and various binders. In the procedure, the surface of the titanium oxide is coated by the other materials to prevent the titanium oxide from separating from the binder owing to the photoinduced reaction. Thus, many efforts have been devoted for a long time to suppressing the photocatalytic activity to avoid chalking. One of the achievements is surface modification (coating) with inorganic materials. Several models of the modification are shown in Figure 1.5A. Coating with an inorganic compound,

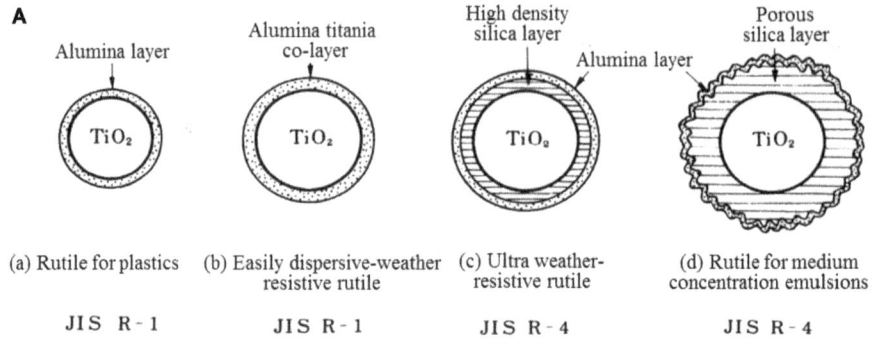

A

Alumina layer

Alumina titania co-layer

High density silica layer

Alumina layer

Porous silica layer

TiO₂   TiO₂   TiO₂   TiO₂

(a) Rutile for plastics   (b) Easily dispersive-weather resistive rutile   (c) Ultra weather-resistive rutile   (d) Rutile for medium concentration emulsions

JIS R-1   JIS R-1   JIS R-4   JIS R-4

B

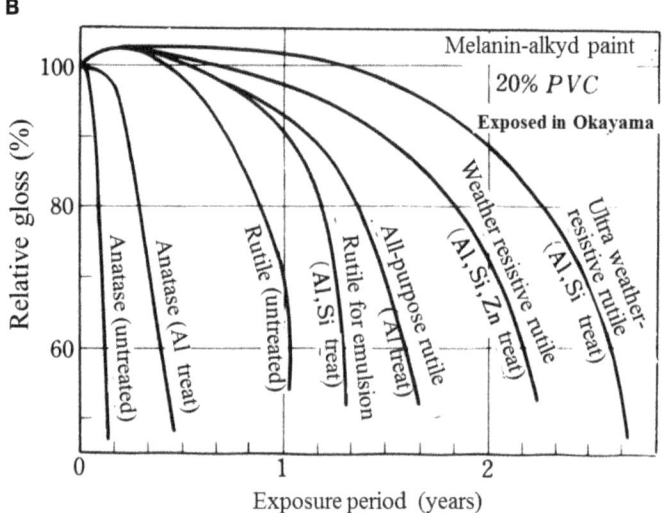

Figure 1.5   (A) Representative surface modifications of TiO₂ with inorganic materials and (B) improvement of the weather resistivity. Reproduced from Manabu Seino, *Sankachitan* (1991) p. 31 and p. 170, with permission from Gihodo Shuppan.

alumina (Al₂O₃), suppresses the catalytic activity on the surface, namely improves the weather resistance as well as the lipophilicity which promotes the dispersibility into the organic solution. On the other hand, surface treatment with silica (SiO₂) promotes the hydrophilicity. Further improvement of weather resistance could be achieved by coating the surface with both alumina and amorphous titanium oxide hydrate.

On the basis of the fabrication procedures, the TiO₂ for pigments could be categorized into the various products suitable for the uses of easily dispersive high-weather-resistant paint, hyper-weather-resistant paint, high concentrated emulsion paint, ink for cans, cardboard veneer, and aqueous slurry.[1] Thus, because the modification of the TiO$_2$ surface with silica or alumina prevents the direct contact between the binder resin and TiO$_2$, chalking can be avoided. The weather resistance of the inactivated titanium oxide was tested by the representative methods measuring the gloss, and the results are shown in Figure 1.5B. When the melamine alkyd paint containing 20% vinyl chloride is exposed to the sunlight, the gloss of non-treated anatase is faded most rapidly. This means that the photocatalytic activity of anatase is considerably high. On the other hand, the photocatalytic activity of rutile is low as compared with that of anatase. Furthermore, it is clearly demonstrated that by several treatments, especially by the surface treatments with alumina or silica, the weather resistance is promoted significantly.

Thus, a lot of efforts have been devoted towards suppressing the photocatalytic activities to increase the weather resistance of TiO$_2$. However, there have been reports for a long time on the researches which employed positively the photocatalytic reactions. Upon the discovery of the Honda-Fujishima effect published around 1970, the principle of the reactions on the surface of titanium oxide under photoirradiation was clarified. Since it opened the possibility to develop the method to decompose water to obtain hydrogen energy, TiO$_2$ attracted much attention. In those days, because of the oil shock as an alternative energy source of oil, solar energy began to gather public attention, which raised the expectation for the potential of TiO$_2$.

1.6   History of Photocatalysis and Related Subjects

Figure 1.6 shows the chronology of the scientific reports for photocatalysis and the related subjects. Since the middle of the 19th century, photoresponsive reactions with uranium salts had been studied to oxidize organic acids, which had been named "photocatalysis" until the beginning of the 20th century.[17] At the beginning of the 20th century, the discoloration by light irradiation was reported to be accelerated on the addition of ZnO to pulp colors.[18] Similar phenomena were observed also for ZnCO$_3$, ZnSiO$_3$, and ZnS, though to a lesser degree. By light irradiation on TiO$_2$ and some other oxides, the formation of colored reaction products in the presence of a specific organic reagent such as glycerin was reported,[48] where the white color of TiO$_2$ changed to blue-grey by light irradiation under reduced pressure while glycerin became dihydroxy-acetone followed by the

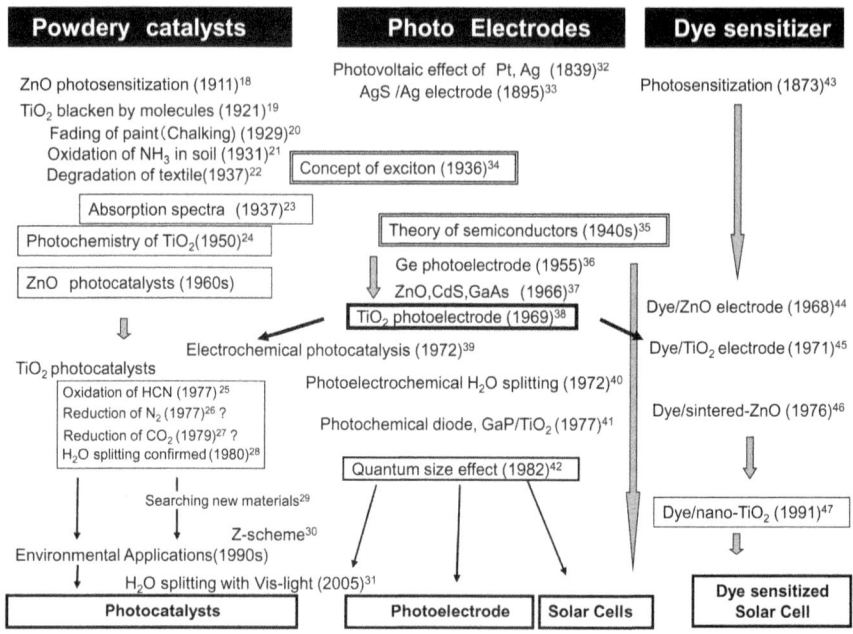

Figure 1.6 The chronology of the reports on photocatalysis and photoelectrochemistry.[18-47]

evolution of CO_2.[19] Metal oxides such as ZnO, PbO, Fe_2O_3, Bi_2O_3 are reported to be blackened in $AgNO_3$ aqueous solution to form silver oxides under the irradiation of strong light.[49] In the same period, it was clarified that the reactions took place at the surface of particulates depending on the electric potential of the substances. The experimental data proved the stoichiometric relation of the products that O_2 was equivalent to 4 Ag.[50] That is, in the anodic process, $4OH^-$ become O_2 and $2H_2O$, while in the cathodic process, Ag^+ becomes Ag.

In 1929, it was reported[51] that TiO_2 paints faded more rapidly than the others when painted films were exposed to the sun and UV light. At that time, the fading of dyes in the presence of TiO_2 was also ascribed to the residual H_2SO_4 in TiO_2.[20] In modern industrialization, one of the most prominent developments is the production of NH_3 as a chemical fertilizer by fixing nitrogen in air. However, when NH_3 fertilizers were used in soil, prolonged exposure to the sunlight caused the formation of nitrite acid. In 1931 this phenomenon was explained by the fact that TiO_2 and ZnO in soil induced the photo-reaction.[21] However, the effect of TiO_2 and ZnO was an issue of argument, because NH_3 itself was known to be decomposed by light irradiation. In the meantime, the effect was verified by the quantitative analysis of the formation of NO_2^- and NO_3^- in TiO_2-suspended $(NH_4)_2CO_3$ aqueous solution by sunlight irradiation.[52] The amount of the reaction products varied significantly by the heat treatments of TiO_2. In the other field of textile technology, the light resistance of man-made fibers was also found to decrease significantly by doping the fiber with TiO_2 powder, which was explained as photosensitization of TiO_2.[22]

In the field of physics, the concept of exciton in the light irradiated crystals was proposed in the 1930s.[34] The fading of dyes with TiO_2 (called "chalking") was explained by the reaction in which the excited state electrons (or exciton) in the pigment moved to the surface to cause oxidation or reduction.[23] The mechanism that potential reactive centers were situated on the interface between dye and TiO_2 particles could be evidenced experimentally. After *"Photochemistry of Rutile"* was compiled in 1950,[24] the photocatalytic action of TiO_2 had not attracted much attention any more, and a photosensitive catalyst ZnO had been mainly used. For TiO_2 photocatalysts, only a few reports were published. For example, the decomposition of tetrarin in solution was reported.[53] In the late 1960s, the electrochemical methods to investigate catalysis by semiconductors were presented.[54]

As for the photo effect on electrodes, in the middle of the 19th century, Becquerel[32] conducted the experiments in which two electrodes were adopted from Pt, Ag, or Pb electrodes and in a dilute acid solution one electrode was irradiated with light of various colors. He noted that (i) the current produced was the greatest for the light at the blue-violet end of the spectrum, (ii) this current was not a thermoelectric effect, and (iii) the illuminated electrode became positive against the dark one. His discovery was followed by numerous investigations. In the meantime, various kinds of photoelectric cells with a light-sensitive electrode which was able to convert radiant energy into electrical energy were developed. Later, Becquerel's observations could be elucidated as a special case of the barrier layer photoelectric cell.[55]

The theory of semiconductors in the field of solid state physics might be initiated with the theory of the photoconductivity of alkali halides crystals.[56] The theory for the surface states and rectification at a metal semiconductor contact was established in the 1940s.[35] Based on these semiconductor physics, the interface between semiconductor single crystal and electrolyte was firstly examined for germanium (Ge),[36] and electrochemical aspects of photoirradiated semiconductor had been developed around 1960.[57] In those days, Ge single crystals were intensively studied and so were single crystals of Si and Se, with a few numbers of reports. In these experiments, the problem came up with the photodissolution of the electrodes. Gerischer[37] considered the case where the "photoelectrochemical" effect is caused by the absorption of light in a semiconducting electrode. However, since he used ZnO single crystal, the electrodes were corroded by the photoirradiation.

For TiO_2 single crystal, which might be rare to obtain with a size large enough to use as an electrode, Fujishima *et al.*[38] reported the result of historic photoelectrochemical experiments in 1969. Different from the other semiconductors, the dark anodic current was extremely low and no dissolution of Ti ions was observed. They confirmed the occurrence of photosensitized electrolytic water oxidation by comparing with the oxidation behavior of I^- ions. The reaction mechanism was interpreted with a band model of a semiconductor proposed by Gerischer.[37]

The effect of organic dyes on the solid materials was firstly reported as an increase in the light sensitivity of the photographic plate by Vogel in 1873.[43] He found that the optical absorption of the substance mixed with silver bromide played important roles. Later, this phenomenon was applied to color photography. The experimental phenomena showed that the energy absorbed by a dye in contact with a solid could transfer to the solid to cause photoelectrical effects. After the development of UV-responsive semiconductor photoelectrodes, the possibility to extend the photoelectrode response to the visible light region became the major concern of research. Photosensitized electrode current has also been investigated in the late 1960s. At first, ZnO single crystal was used as the substrate electrode to which electron transfer from photoexcited dye adsorbed on the surface took place.[44] The oxidized dye may either be reduced by some reactant in solution, or it might oxidize the surface of ZnO. Then, it was suggested that the electron should transfer from the excited dye to the ZnO solid. Since TiO_2 single crystal was proved to be stable in water, TiO_2 has been used in place of ZnO.[45]

Researches on photocatalysis by TiO_2 gathered attention again in 1972 since the high stability was revealed by the historic report on the electrochemical water splitting by use of TiO_2 single crystals.[40] Figure 1.7 shows the number of documents that appeared in SciFinder of the American Chemical Society which contain the key words "TiO_2" and "photocatalysis". Though the number of patents became constant in the late 1990s, the research documents increased exponentially from the beginning of 1970. As is seen in Figure 1.7, many basic researches on photocatalysis have been published from the 1970s to 1980s. In the same year that the photoelectrochemical study

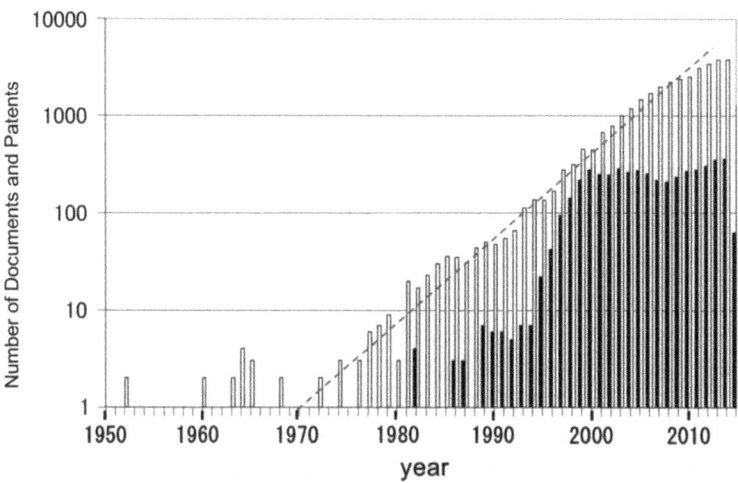

Figure 1.7 The number of scientific reports (white bar) and patents (black bar) of TiO_2 photocatalysis based on the data of SciFinder (American Chemical Society).

of TiO_2 by Fujishima-Honda was published, it was demonstrated that the electrochemical analysis based on the local cell process facilitated the estimation of a rate of the photocatalytic reaction of organic molecules.[39] Then, several applications of TiO_2 photocatalytic reactions were successively reported, for example, photodetoxification of HCN,[25] NH_3 formation from N_2 and water,[26] reduction of CO_2,[27] and H_2 production from carbohydrate and water.[58] Among them, the report in 1977 that NH_3 and hydrazine (N_2H_4) were generated on light irradiation over the wet titanium oxide containing Fe ion under N_2 atmosphere[26] gathered attention. However, such suspicion that TiO_2 surface was probably not so unique to be able to participate in the photocatalytic fixation of N_2 arose.[59] Despite the large number of reports, none of the suggestive reactions has been proposed unequivocally. Since several reports which were suspicious about the results appeared, it was regarded that the experiments were not correct.[60] As for the photocatalytic reduction of CO_2 with UV light, the detection of produced O_2 in the early researches may be doubtful. Quite recently, the reduction of CO_2 to CO with the formation of H_2 and O_2 was accomplished using a sophisticated photocatalyst, Ag/$BaLa_4Ti_4O_{15}$ nanocrystallite,[61] which had been developed for solar water splitting.

Photocatalytic decomposition of H_2O into H_2 and O_2, *i.e.*, water splitting, is also a difficult reaction to attain because it is a highly endoenergetic (up-hill) reaction. Upon the discovery of the

Honda-Fujishima effect, as described above the principle of the reactions on the surface of TiO_2 under the photoirradiation was clarified. Since it opened the possibility to develop the method to
decompose water to obtain hydrogen energy, TiO_2 won attention. In
those days, because of the oil shock, as an alternative energy source of
oil, solar energy began to gather public attention, which raised the
expectation for TiO_2. In the early reports for the photocatalytic formation of H_2 from water, the oxidation of water was not verified
evidently. The contaminations might be oxidized in place of H_2O. On
the other hand, in the gas phase the photocatalytic evolution of both
O_2 and H_2 was not confirmed until 1980, where Pt deposited TiO_2
powder was employed.[28] This photocatalyst has been developed as a
small, short-circuited Pt–TiO_2 particulate electrode system,[62] which
showed a mechanism similar to that of the Schottky type photochemical diode.[63]

Figure 1.8 shows several types of photocatalysts originated from
the particulate electrodes suggested in the late 1970s.[64] In the figure,
(a) and (b) show the photocatalyst models in which n- and p-type
semiconductor particles were suspended in water. As shown in
Figure 1.8(a) for n-type semiconductor, excitation takes place at the

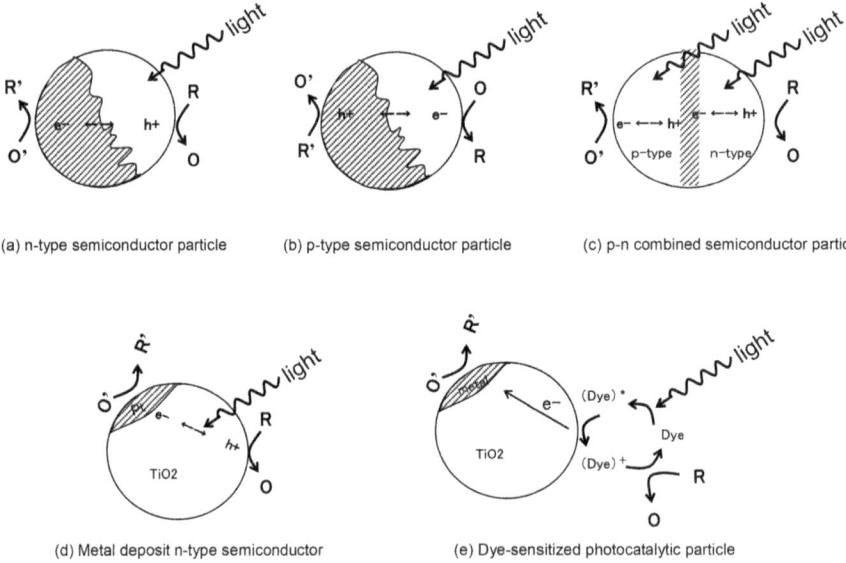

(a) n-type semiconductor particle   (b) p-type semiconductor particle   (c) p-n combined semiconductor particle

(d) Metal deposit n-type semiconductor   (e) Dye-sensitized photocatalytic particle

Figure 1.8 Several possibilities of photoelectrochemical reactions using
semiconductor particles.
Reproduced from A. J. Bard, Photoelectrochemistry and heterogeneous photocatalysis at semiconductors, *J. Photochem.*,
10, 59–75. Copyright (1979), with permission from Elsevier.

irradiated part in the semiconductor particle and then the formed electron moves to the non-irradiated part. Thus oxidation and reduction take place at the irradiated surface and the shadow surface, respectively. In this figure, O and O′ represent the oxidized form, and R and R′ represent the reduced form of the reactant, respectively. Compared to the n-type, for the p-type semiconductor, reverse reactions take place as shown in Figure 1.8(b); holes generated at the irradiated part move to the non-irradiated part to cause an oxidation reaction, while electrons reduce the reactant at the irradiated surface. Since this model originates from the electrode reaction, it does not actually seem to work. The diameter of the powder is usually as small as that for the light penetration depth to pass throughout, and the scattered light may shine on the "backside" of the particle. Model (c) originates from the photochemical diode, in which n-type and p-type semiconductor electrodes were combined with each other with ohmic contact (the electronic state has no barrier and electrons can flow freely between the semiconductors as will be described in Chapter 7). In this model the oxidation and reduction take place at the n-type and the p-type part of the combined semiconductor, respectively. It may have some reality but it would not be easy to make the excitation rates for both semiconductors equalized. Model (d) is based on the electrode cell used to demonstrate the Honda-Fujishima effect. Namely, TiO_2 semiconductor is connected directly to Pt metal, and water molecules are reduced to H_2 and oxidized to O_2 at Pt and TiO_2 surfaces, respectively. In order to utilize visible light energy, dye may be used as photosensitizer as shown in model (e), where the excited dye releases electrons to the semiconductor and the surface-deposited metal receives the electrons to reduce the substances.

In the 1980s, it was realized that TiO_2 did not absorb visible light and was not suitable for the utilization of solar energy. Then, the research trend shifted to the visible-light responsive photocatalysts such as CdS and many materials have been tested as candidates for water splitting.[65] It was only in 2005 that a stable composite semiconductor was developed for water splitting by visible light.[31]

In the 1990s, the practical applications by utilizing the oxidation and reduction powers of TiO_2 were initiated, aiming at the decomposition of harmful organic materials of a very small amount. Later, it was recognized that the decomposition of small amounts of organic compound could be applied to many fields such as environmental clean-up. Then, photocatalysis of TiO_2 gathered considerable attention for the decomposition of small harmful materials.

At present, the modification of TiO_2 is in progress, aiming at the achievement of higher activity with visible-light. In addition to the decomposition of organic materials, a function of superhydrophilicity of TiO_2 was discovered,[66] which is now practically used for mirrors and window glasses.

References

1. M. Seino, *Titanium Dioxide*, Gihohdo (in Japanese), 1991.
2. *Landolt-Börnstein, Numerical Data and Functional Relationships in Science and Technology*, ed. O. Madelung, Springer-Verlag, 1984, group III, vol. 17g, section 9.15.2.1.
3. M. Voinov, J. Augustynski, in *Heterogeneous Photocatalysis*, ed. M. Schiavello, John Wiley & Sons, 1997.
4. JCPDS (Joint Committee on Powder Diffraction Standards) Card #21-1276.
5. JCPDS, Card #21-1272.
6. JCPDS, Card #36-1451.
7. L. Kavan *et al.*, *J. Am. Chem. Soc.*, 1996, **118**, 6716.
8. K. I. Hadjiivanov and D. G. Klissurski, *Chem. Soc. Rev.*, 1996, **25**, 61.
9. G. Pacchioni, A. M. Ferrari and P. S. Bagus, *Surface Sci.*, 1996, **350**, 159.
10. H. Kominami and B. Ohtani, *Electrochemistry*, 1998, **66**, 996.
11. M. J. Gázquez *et al.*, *Mater. Sci. Appl.*, 2014, **5**, 441.
12. *Handbook of Chemistry and Physics*, ed. D. R. Lide, CRC Press, 2002.
13. A. Berge, *Pottery Gaz. Glass Trade Rev.*, 1913, **38**, 1052 (*Chem. Abst.* 7, 25461).
14. N. Heaton, *J. R. Soc. Arts*, 1922, **70**, 552 (*Chem. Abst.* **16**, 17458).
15. S. P. Pappas and R. M. Fischer, *J. Paint Technol.*, 1974, **46**, 65.
16. H. G. Voelz, G. Kaempf and A. Klaeren, *Farbe und Lack*, 1976, **82**, 805.
17. H. Fay, *Am. Chem. J.*, 1896, **18**, 269.
18. A. Eibner, *Chem.–Ztg.*, 1911, **35**, 753, 774, 786b (*Chem. Abst. 5*, 21155, 22909).
19. C. Renz, *Helv. Chim. Acta*, 1921, **4**, 961.
20. H. Wagner, *Farben-Ztg.*, 1929, **34**, 1243 (*Chem. Abst. 23*, 24035).
21. G. G. Rao and N. R. Dhar, *Soil Sci.*, 1931, **31**, 379.
22. M. Horio, *Nippon Gakujutsu Kyokai Houkoku*, 1937, **12**, 204.
23. C. F. Goodeve, *Trans. Faraday Soc.*, 1937, **33**, 340.

24. W. A. Weyl and T. Foerland, *Ind. Eng. Chem.*, 1950, **42**, 257.
25. S. N. Frank and A. J. Bard, *J. Am. Chem. Soc.*, 1977, **99**, 303.
26. G. N. Schrauzer and T. D. Guth, *J. Am. Chem. Soc.*, 1977, **99**, 7189.
27. T. Inoue, A. Fujishima, *et al.*, *Nature*, 1979, **277**, 637.
28. S. Sato and J. M. White, *Chem. Phys. Lett.*, 1980, **72**, 83.
29. A. Kudo and Y. Miseki, *Chem. Soc. Rev.*, 2009, **38**, 253.
30. K. Sayama, H. Arakawa, *et al.*, *Chem. Commun.*, 2001, 2416.
31. K. Maeda, K. Domen, *et al.*, *J. Am. Chem. Soc.*, 2005, **127**, 8286.
32. E. Becquerel, *Compt. Rend.*, 1839, **9**, 561.
33. A. W. Copeland, O. D. Black and A. B. Garrett, *Chem. Rev.*, 1942, **31**, 177.
34. Y. Frenkel, *Phys. Z. Sowjetunion*, 1936, **9**, 158.
35. J. Bardeen, *Phys. Rev.*, 1947, **71**, 717.
36. W. H. Brattain and C. G. B. Garrett, *Bell Syst. Tech. J.*, 1955, **34**, 129.
37. H. Gerischer, *J. Electrochem. Soc.*, 1966, **113**, 1174.
38. A. Fujishima, K. Honda and S. Kikuchi, *Kogyo Kagaku Zasshi*, 1969, **72**, 108.
39. H. Yoneyama, Y. Toyoguchi and H. Tamura, *J. Phys. Chem.*, 1972, **76**, 3460.
40. A. Fujishima and K. Honda, *Nature*, 1972, **238**, 37.
41. A. J. Nozik, *Appl. Phys. Lett.*, 1977, **30**, 567.
42. Al. L. Efros and A. L. Efros, *Sov. Phys. Semicond.*, 1982, **16**, 772.
43. H. Vogel, *Ber. Dtsch. Chem. Ges.*, 1873, **6**, 1302.
44. H. Gerischer and H. Tributsch, *Ber. Bunsenges. Phys. Chem.*, 1968, **72**, 637.
45. K. Honda, *J. Photochem. Photobiol., A*, 2004, **166**, 63.
46. H. Tsubomura *et al.*, *Nature*, 1976, **261**, 402.
47. B. O'Regan and M. Graetzel, *Nature*, 1991, **353**, 737.
48. C. Renz, *Z. Anorg. Allg. Chem.*, 1920, **110**, 104 (*Chem. Abst.* **14**, 18116).
49. G. Tammann, *Z. Anorg. Allg. Chem.*, 1920, **114**, 151.
50. E. Baur, *Helv. Chim. Acta*, 1924, **7**, 910.
51. E. Keidel, *Farben-Ztg.*, 1929, **34**, 1242 (*Chem. Abst.* **23**, 24034).
52. S. Osugi and M. Aoki, *Nippon Dojo Hiryogaku Zasshi*, 1936, **10**, 11.
53. S. Kato and F. Mauo, *Kogyo Kagaku Zasshi*, 1964, **67**, 1136.
54. T. Freund and W. P. Gomes, *Catal. Rev.*, 1969, **3**, 1.
55. R. H. Mueller and A. Spector, *Phys. Rev.*, 1932, **41**, 371.
56. A. L. Hughes, *Rev. Mod. Phys.*, 1936, **8**, 294.
57. H. Gerischer, *Adv. Electrochem. Electrochem. Eng.*, 1961, **1**, 139.
58. T. Kawai and T. Sakata, *Nature*, 1980, **286**, 474.
59. R. I. Bickley and V. Vishwanathan, *Nature*, 1979, **280**, 306.

60. J. A. Davies, D. L. Boucher and J. G. Edwards, *Adv. Photochem.*, 1995, **19**, 235.
61. K. Iizuka *et al.*, *J. Am. Chem. Soc.*, 2011, **133**, 20863.
62. B. Kraeutler and A. J. Bard, *J. Am. Chem. Soc.*, 1978, **100**, 2239.
63. A. J. Nozik, *Appl. Phys. Lett.*, 1976, **29**, 150.
64. A. J. Bard, *J. Photochem.*, 1979, **10**, 59.
65. A. Fujishima, T. N. Rao and D. A. Tryk, *J. Photochem. Photobiol., C*, 2000, **1**, 1.
66. R. Wang, A. Fujishima, *et al.*, *Nature*, 1997, **388**, 431.

2 Principles of Light Over Solids

2.1 Intensity and Absorption of Light

Light is an electromagnetic wave which can be discriminated by wavelength, and possesses the property as a particle as well. The light of the wavelength of λ, or frequency of $\nu = c/\lambda$, is the assembly of the particles with the energy of ch/λ, or $h\nu$, where c and h are the speed of light ($c = 3.00 \times 10^8$ ms^{-1}) and Planck constant ($h = 6.63 \times 10^{-34}$ Js), respectively. Because energy familiar in daily life is electricity, it would be easier to understand by expressing energy in terms of electric voltage. Namely, by dividing the energy in J (joule) unit by the electric charge of an electron (elementary charge $e = 1.60 \times 10^{-19}$ C), it can be transferred to the electric voltage in V (volt) unit. It is generally convenient to express energy in the unit of electron volt (eV). Then the light of wavelength λ (nm) possesses the energy E as expressed by eqn (2.1),

$$E \text{ (eV)} = 1240/\lambda \tag{2.1}$$

where the coefficient 1240 is calculated from $ch/(10^{-9} e)$.

Electromagnetic waves span from radio waves to ionizing radiation. Figure 2.1 shows the relationship between wavelength, frequency, energy, and temperature for a wide range of electromagnetic waves. Here, the temperature of the corresponding energy was calculated from energy divided by the Boltzmann constant, $k_B = 1.38 \times 10^{-23}$ J K^{-1}. The light to which human eyesight responds is called visible light. The wavelength of visible light ranges from 400 nm to 700 nm. Light with shorter wavelength is called ultraviolet (UV) light, and that with longer wavelength is called infrared (IR) light.

Introduction to Photocatalysis: From Basic Science to Applications
By Yoshio Nosaka and Atsuko Nosaka
© Yoshio Nosaka and Atsuko Nosaka, 2016
Published by the Royal Society of Chemistry, www.rsc.org

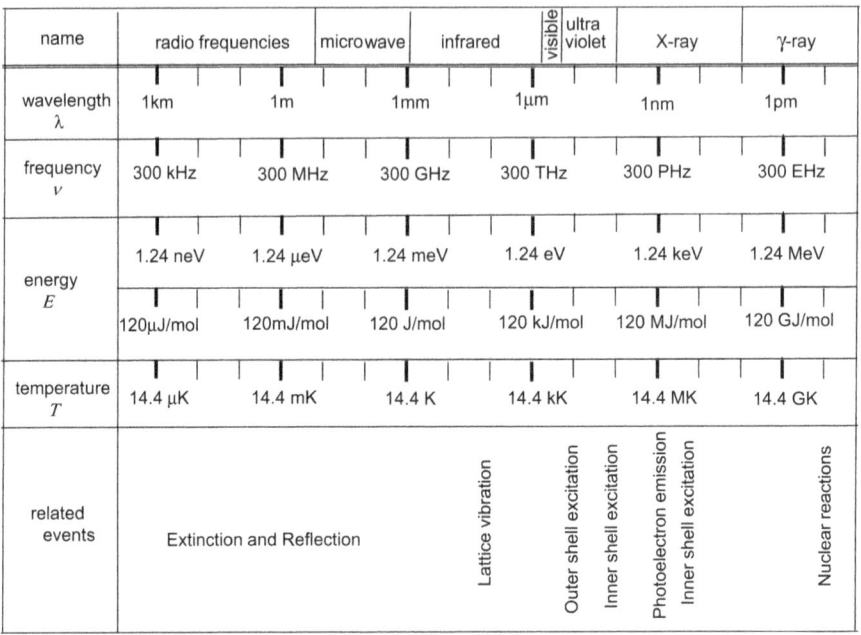

name	radio frequencies	microwave	infrared	visible	ultra violet	X-ray	γ-ray
wavelength λ	1km 1m		1mm	1µm		1nm	1pm
frequency ν	300 kHz 300 MHz		300 GHz	300 THz		300 PHz	300 EHz
energy E	1.24 neV 1.24 µeV		1.24 meV	1.24 eV		1.24 keV	1.24 MeV
	120µJ/mol 120mJ/mol		120 J/mol	120 kJ/mol		120 MJ/mol	120 GJ/mol
temperature T	14.4 µK 14.4 mK		14.4 K	14.4 kK		14.4 MK	14.4 GK
related events	Extinction and Reflection			Lattice vibration	Outer shell excitation / Inner shell excitation	Photoelectron emission / Inner shell excitation	Nuclear reactions

Figure 2.1 Various electromagnetic waves and the related events on substances.

The events that occur by the interaction of electromagnetic waves with materials are also shown in Figure 2.1.

The intensity of light is often expressed by I and is easily measured with an optical power meter. The unit is expressed by the power of unit area, $W\,m^{-2}$. Power does not mean energy itself but corresponds to the energy per second ($W = J\,s^{-1}$). Therefore, the peak power, which shows the instantaneous power, and pulse energy, which describes the energy during a short period, exist, but there can be no concept of pulse power and peak energy in scientific meaning. The intensity of pulse laser is measured with a power meter as the light energy averaged over 1 s. Then, the energy of one light pulse can be obtained by dividing the power by the number of pulses per second. The peak power of the pulse can be estimated by dividing the pulse energy by the pulse width (or duration of the light pulse).

The number of photons is important for photoinduced reactions. When light of intensity I ($mW\,cm^{-2}$) with wavelength λ (nm) propagates into the cross section of $1\,cm^2$, the number of the incident photons becomes $10^{-12}\,\lambda I/ch$, which can be calculated to be $5.0 \times 10^{12}\,\lambda I$ (photons $cm^{-2}\,s^{-1}$) by employing the constant values of c and h. By taking into account that 1 mol is 6.02×10^{23}, the light

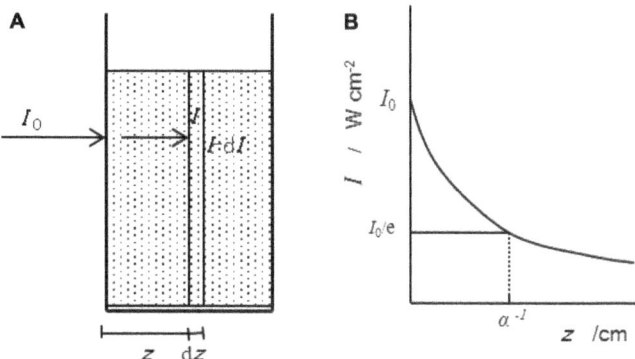

Figure 2.2 Decrease of light intensity (*I*) as a function of penetration distance (*z*).

intensity I (mW cm^{-2}) with wavelength λ (nm) can be described as the photons in the mol unit to be 8.36×10^{-12} λI (mol cm^{-2} s^{-1}). To describe the photon number of 1 mol, the unit of Einstein (Ein) is sometimes used.

Figure 2.2 shows how the light is absorbed in a material or solution. When the light of intensity I advances into the substance, the minute intensity of dI is absorbed within a minute distance dz. Then, the following relationship holds,

$$-\frac{\mathrm{d}I}{\mathrm{d}z} = \alpha I \tag{2.2}$$

where α is an absorption coefficient whose unit is cm^{-1} in most cases. Taking the intensity I_0 at the incident position ($z = 0$), this differential equation can be solved as follows.

$$I = I_0 e^{-\alpha z} \tag{2.3}$$

Because absorbance (Abs) is indicated by a common logarithm on a spectrometer, the absorbance for the material whose thickness and absorption coefficients are d and α, respectively, becomes

$$\mathrm{Abs} = \log\left(\frac{I_0}{I}\right) = 0.4343\ \alpha d \tag{2.4}$$

The relationship of eqn (2.4) is known as the Lambert–Beer rule.

By the way, the Abs of the solution containing molecules of concentration c (M) and molar absorption coefficient ε (cm^{-1} M^{-1}) is described as $\varepsilon c d$, where M is the unit of the molar concentration (M $=$ mol dm^{-3}). The value of εc or ($0.4343\ \alpha$) is sometimes called optical density (OD).

Let's think about the decay of the light intensity, eqn (2.3), from the viewpoint of the photons described above. When the light of wavelength λ (nm) and intensity I_0 (mW cm^{-2}) advances into the substance of absorption coefficient α (cm^{-1}), the density of photons absorbed dI/dz at the position immediately after the incidence ($z=0$) is calculated with eqn (2.2) to be αI_0 (mW cm^{-3}) $= 8.36\times10^{-9}\lambda\alpha I_0$ (M s^{-1}). And as the light advances at the position z, according to eqn (2.3), the density of the absorbed photon decreases by the ratio of $e^{-\alpha z}$. Namely, by substituting eqn (2.3) for eqn (2.2), eqn (2.5), which gives the density of the absorbed photons, is obtained.

$$\frac{\mathrm{d}I}{\mathrm{d}z} = \alpha I_0 e^{-\alpha z} \qquad (2.5)$$

Then let's try to calculate the photon absorption of sunlight, which has intensity of about 1 mW cm^{-2} with peak wavelength around 500 nm. Figure 2.3 shows the change of the photon density when the absorption coefficient α changes from $\alpha_1 = 0.2303$ to $\alpha_4 = 230.3$ cm^{-1}. For the substance of absorption coefficient $\alpha = 2.303$ cm^{-1} (or OD $= 1.0$ cm^{-1}) at the wavelength of 500 nm, the penetration depth (defined by the distance at which the light intensity becomes $1/e$) is 0.4343 cm, and the distance at which the light intensity becomes 1/10 is 1.0 cm. The concentration of the absorbed light near the incidence per second is 1.0×10^{-5} Ms^{-1} and it becomes 1/10, namely 1.0×10^{-6} Ms^{-1} at the position of 1 cm progress. However, for the substance whose absorbance is 100 times larger, namely $\alpha = 230.3$ cm^{-1},

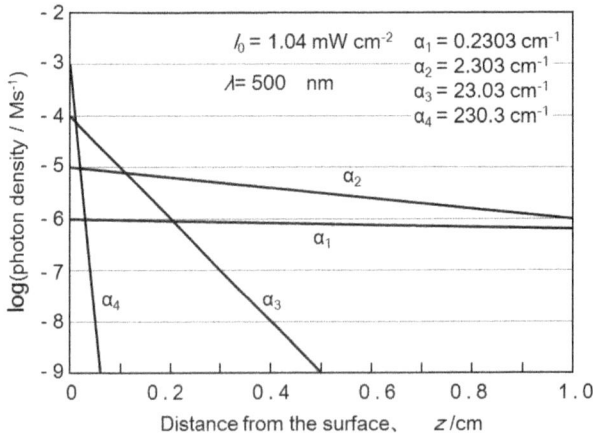

Figure 2.3   Changes of the light density in the material when light of wavelength (λ) of 500 nm and intensity (I_0) of 1.04 mW cm^{-2} is absorbed in the material with different absorption coefficients, from α_1 to α_4.

the penetration depth becomes 43.43 μm. Therefore the light is mostly absorbed near the surface. The photoabsorption density near the incidence becomes 100 times larger in proportion with α, namely 1.0×10^{-3} Ms^{-1} but it becomes nearly 0 at the position of 1 cm progress.

2.2 Optical Constants and Dielectric Constant

Light is an electromagnetic wave. Therefore, according to Maxwell's equation, light imparts an electric field (E) and a magnetic field (H) to the electrons in the substance having electrical conductivity σ as shown in Figure 2.4.

Electric flux density, D, and magnetic flux density, B, can be expressed with E and H by the following eqn (2.6) and (2.7), respectively,

$$D = \varepsilon\, E = \varepsilon_0 \varepsilon_1\, E \tag{2.6}$$

$$B = \mu\, H = \mu_0 \mu_1\, H \tag{2.7}$$

where ε and ε_0 are dielectric constant (permittivity) of substance and vacuum, respectively, and μ and μ_0 are permeability of substance and vacuum, respectively. Usually, $\mu_1 = 1$ holds because common substances are non-magnetic. From Maxwell's equation, using some vector mathematics, one can obtain the following differential equation.

$$\nabla^2 E - \sigma \mu_0 \frac{\partial E}{\partial t} - \varepsilon_0 \varepsilon_1 \mu_0 \frac{\partial^2 E}{\partial t^2} = 0 \tag{2.8}$$

The representative solution of eqn (2.8) is

$$E_x = E_o e^{-i\omega\left(t - \frac{z}{v}\right)} \tag{2.9}$$

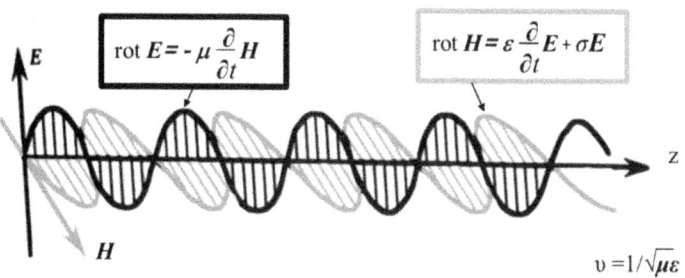

Figure 2.4 Electromagnetic wave depicted by the vectors of electric field E and magnetic field H according to Maxwell's equation.

with

$$\frac{1}{v} = \sqrt{\mu_0 \left(\varepsilon_0 \varepsilon_1 + i\frac{\sigma}{\omega}\right)} \tag{2.10}$$

That is, the electric field of the x direction, E_x, formed is a sine wave with angular frequency ω and moving to the z direction at the velocity of v. If we define complex permittivity, ε^* $(=\varepsilon_1 + i\varepsilon_2)$, eqn (2.10) becomes

$$\frac{1}{v} = \sqrt{\mu_0 \varepsilon_0 \varepsilon^*} \tag{2.11}$$

with

$$\varepsilon_2 = \frac{\sigma}{\omega \varepsilon_0} \tag{2.12}$$

Furthermore, if we define complex refractive index n^* $(=n + ik)$ as the square root of ε^*,

$$n^* = \sqrt{\varepsilon^*} \tag{2.13}$$

Eqn (2.11) becomes eqn (2.14) because the speed of light in a vacuum, c, is described as $c = 1/\sqrt{\varepsilon_0 \mu_0}$. By using this relationship, the value of ε_0 is defined as $\varepsilon_0 = \mu_0^{-1} c^{-2}$ $(= 8.85 \times 10^{-12}$ Fm^{-1} because $\mu_0 = 4\pi \times 10^{-7}$ Hm$^{-1})$.

$$\frac{1}{v} = \frac{n^*}{c} = \frac{n + ik}{c} \tag{2.14}$$

By substituting eqn (2.14) for eqn (2.9), the representative solution of Maxwell's equation becomes eqn (2.15).

$$E_x = E_0 e^{-i\omega\left(t - \frac{nz}{c}\right)} e^{-\left(\frac{\omega k}{c}\right)z} \tag{2.15}$$

The second exponential function in eqn (2.15) indicates that an oscillating electric field $E_0 e^{-i\omega t}$ is decreased at the rate of $\left(\dfrac{\omega k}{c}\right)$ when it advances to the z direction. Because the light intensity I is proportional to the square of the amplitude of electric field E_x, eqn (2.16) is given.

$$I = |E_x|^2 \propto e^{-\alpha z} \tag{2.16}$$

Thus, the absorption coefficient α is given with k (the imaginary part of n^*) as expressed by eqn (2.17).

$$\alpha = \frac{2\omega k}{c} \tag{2.17}$$

From this correlation, k is called the "absorption index". Angular frequency ω ($\omega = 2\pi\nu$) is expressed by using the wavelength in a vacuum, λ_{vac}, as $\omega = 2\pi c/\lambda_{vac}$. Because the refractive index n of air is about 1.0003, the wavelength in the air λ ($= \lambda_{vac}/n$) is almost equal to λ_{vac}. Taking $\alpha = 4\pi k/\lambda$, when k is equal to 0.080 ($= 1/4\pi$), α becomes $1/\lambda$, *i.e.*, the penetration depth ($= 1/\alpha$) becomes equal to the wavelength of the light.

The relationship between ε^* and n^* of eqn (2.13) is mathematically equivalent to the correlations given by eqn (2.18)–(2.21) for the components.

$$\varepsilon_1 = n^2 - k^2 \tag{2.18}$$

$$\varepsilon_2 = 2nk \left(= \frac{\sigma}{\omega\varepsilon_0} \right) \tag{2.19}$$

$$n = \left(\frac{1}{2} \left(\sqrt{\varepsilon_1^2 + \varepsilon_2^2} + \varepsilon_1 \right) \right)^{\frac{1}{2}} \tag{2.20}$$

$$k = \left(\frac{1}{2} \left(\sqrt{\varepsilon_1^2 + \varepsilon_2^2} - \varepsilon_1 \right) \right)^{\frac{1}{2}} \tag{2.21}$$

If one compares eqn (2.19) with eqn (2.12), the relation between the absorption coefficient α and the electrical conductivity σ or ε_2 (imaginary part of the permittivity) can be obtained as eqn (2.22).

$$\alpha = \frac{\sigma}{c\varepsilon_0 n} = \frac{\omega\varepsilon_2}{cn} \tag{2.22}$$

This equation indicates that substances having a higher conductivity σ, such as metals, usually show higher absorbance, which means that they are not transparent.

It should be noted that "refractive index" originates from the fact that the velocity of light decreases by the factor of n owing to the electron's motion caused by the oscillating electromagnetic field. The refraction of light is one of the familiar evidences of the decrease of the velocity. Actually, when light is incident vertically to a substance, no refraction is observed but the velocity is definitely reduced by the factor of the refractive index, n.

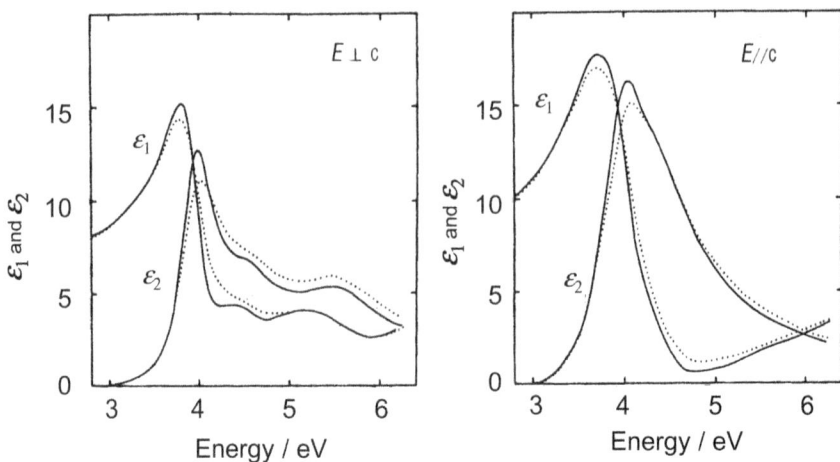

Figure 2.5   The dielectric constants of rutile TiO$_2$. The solid and dotted lines correspond to the values measured at the temperatures of 298 and 0 K, respectively.
Reproduced from K. Vos and H. J. Krusemeyer, Reflectance and electroreflectance of TiO$_2$ single crystals: I. Optical spectra, *J. Phys. C: Solid State Phys.*, **10**, 3893–3939. Copyright (1977), with permission from IOP publishing.

Now let's see the experimental data measured for TiO$_2$. Figure 2.5 shows the permittivity (dielectric constant) of rutile TiO$_2$.[1] Because the rutile crystal shows dual refraction, the set of dielectric constants, ε_1 and ε_2, changes depending on the direction of the crystal axis c, which is parallel or perpendicular to the electric field E of the polarized light. In Figure 2.6A, the logarithm of the absorption which is calculated with reflectance and transmittance at various photon energies is plotted,[2] and in Figure 2.6B the components of the complex refractive index, n and k, of amorphous TiO$_2$ are shown.[3] The horizontal axis in the figure is expressed in energy units of the light, which can be easily transformed to the wavelength by using eqn (2.1).

As was seen in Figure 2.5, the real part ε_1 of ε^* is correlated with the imaginary part ε_2. And the real part n and imaginary part k of the complex optical constants in Figure 2.6B correlate with each other. This bidirectional property, which is called the Kramers–Kronig relationship, originates from the mathematical properties of complex analysis. This relationship connects the real and imaginary parts of any complex function, and it is often used to calculate the real part from the imaginary part (or *vice versa*) of response functions. For example, the wavelength dependence of n could be calculated from that of k or α.

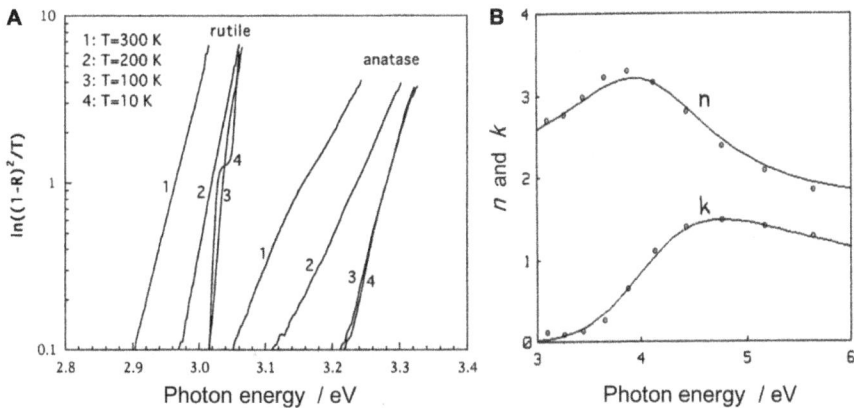

Figure 2.6 (A) Optical onset plotted in a log scale *versus* photon energy for anatase and rutile crystals at different sample temperatures. *R* and *T* signify reflectance and transmittance, respectively. Reprinted from H. Tang, H. Berger, P. E. Schmid and F. Levy, Optical properties of anatase (TiO_2), *Solid State Commun.*, **92**, 267–271, Copyright (1994), with permission from Elsevier. (B) The refractive index *n* and extinction coefficient *k* of amorphous TiO_2 *versus* photon energy. Reprinted (figure) with permission from A. R. Forouhi and I. Bloomer, *Phys. Rev. B: Condens. Matter Mater. Phys.*, **34**, 7018. Copyright (1986) by the American Physical Society.

2.3 Reflection and Interference of Light

When light propagates into a substance with a complex refractive index of n^* from a medium with a refractive index of n_m, some portion (r) of the light is reflected. Because the wavelength of light reduces by the factor of n/n_m with the decrease of the light velocity, the height of E_0 must be reduced by the factor of $1 - r$ to keep the smooth connection of the trigonometric function of E_x from the medium to the substance at the surface of the substance (as shown in Figure 2.7). The electric field intensity before propagation of light on the surface should be the sum of the reflection part, r, in the medium and the propagation part, $(1 - r)$, in the substance. Thus, one suspects that eqn (2.23) can be derived from Figure 2.7.

$$n_m E_0 = -n_m\, r E_0 + n^*(1 - r)E_0 \tag{2.23}$$

This gives

$$r = \frac{n^* - n_m}{n^* + n_m} \tag{2.24}$$

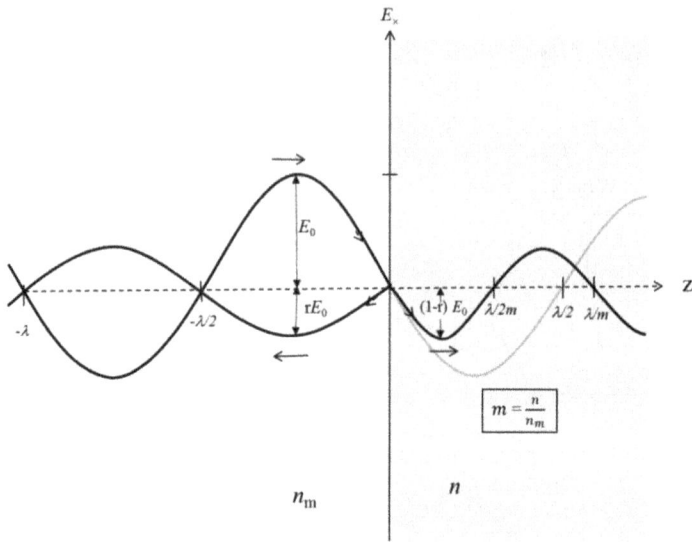

Figure 2.7   Illustration of the reflection at the interface where an electromagnetic wave from medium with refractive index n_m hits the
surface of a substance with refractive index n to be reflected
with the fraction r.

Since $n^* = n + ik$, for the light of intensity I, the ratio of the reflection
R can be expressed by

$$R = |r|^2 = \frac{(n-n_m)^2 + k^2}{(n+n_m)^2 + k^2} \tag{2.25}$$

Light is reflected both on incidence into the material with a high
refractive index from the medium (like air) with a low refractive index,
and on coming out from the material into the medium. Glass does
not absorb visible light $(k=0)$ but, since the refractive index n is 1.5
and that of air n_m is 1.0, from eqn (2.25), 4% of the light is calculated
to be reflected both at the inlet and the outlet when passing through
glass. Therefore, a total amount of about 8% of the light is reflected,
while 92% of the light is transmitted through the glass. Namely, even
for a transparent material, when the refractive index is different from
that of air as is often the case, a part of the light does not propagate
but is reflected. This is the reason why moonlight is reflected on a
water surface at night. It might be familiar that a glass or a lens is
coated with stacked thin films with different refractive indexes to
prevent the reflection on a glass surface.

Eqn (2.25) shows that R becomes close to unity when k becomes
large and eqn (2.19) shows that substances with large conductivity σ

such as metals have large k, which indicates that metals have the property of light reflection.

When the thickness of a glass through which light passes becomes close to the wavelength of the light, the reflective light and the incident light overlap at the outlet. Interference for the light of wavelength λ takes place at the film thickness which becomes a multiple of $\lambda/2n$. This is the reason why an oil film on water and a water film of soap bubbles appear iridescent with rainbow colors. The absorption spectra of TiO_2 thin films present periodically the maximum and the minimum due to the interference as shown in Figure 2.8.[4]

Virtual absorbance by interference can be given with the following equation, where d is the film thickness, and the parameters c_1 and c_2 are the functions of refractive indexes.

$$\text{Abs} = \log\left(1 + c_1 \cos\left(\frac{4\pi n d}{\lambda}\right)\right) + c_2 \qquad (2.26)$$

In this case, for the known refractive index n of the film, the film thickness can be roughly estimated with the following equation, by taking the wavelengths at maximum λ_1, and the minimum λ_2.

$$d = \frac{\lambda_1 \lambda_2}{4n|\lambda_1 - \lambda_2|} \qquad (2.27)$$

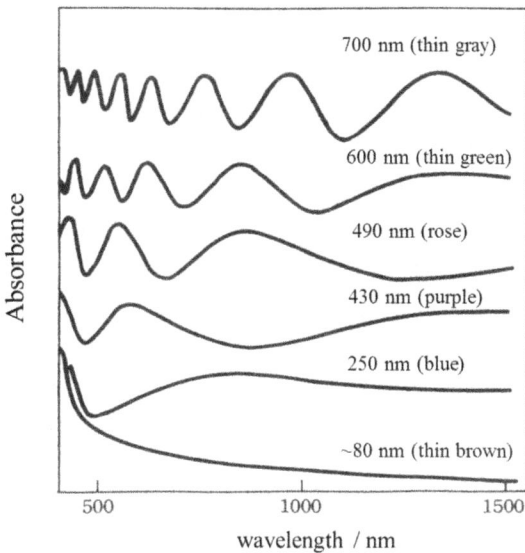

Figure 2.8 Absorption spectra of TiO_2 coating films prepared with a sol–gel method. Thickness and apparent colors are shown in the figure. Reproduced from *Zoru-Geru Hou no Kagaku* (1988) with permission from Agune-Shofusha.

2.4   Extinction of Light by Particles

Extinction of light by particles is caused by scattering and absorption. Scattering takes place when light hits the materials of heterogeneous shape through a medium with a different refractive index. When N spherical particles with the radius of a are in the unit volume $(1\,\text{cm}^3)$, the apparent extinction coefficient γ is expressed as follows.

$$\gamma = N\pi a^2 Q_{\text{ext}} \tag{2.28}$$

The extinction factor of light, Q_{ext}, is described as the sum of the extinctions for scattering Q_{sca}, and absorption Q_{abs}.[5]

$$Q_{\text{ext}} = Q_{\text{sca}} + Q_{\text{abs}} \tag{2.29}$$

To calculate Q of the particle with any radius and complex refractive index, Mie theory must be employed. However, this calculation is quite complicated.[5,6] Some parts of the results calculated by the Mie theory are shown in Figures 2.9 and 2.10. For convenience, a

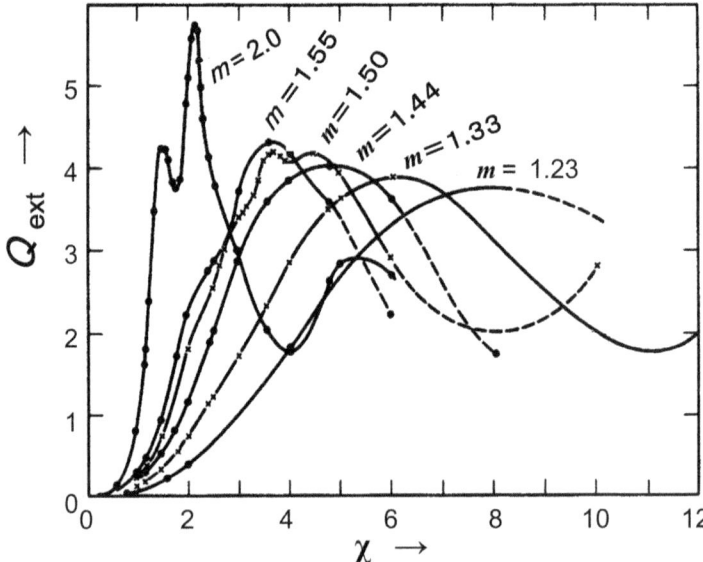

Figure 2.9   The value of Q_{ext} calculated from Mie theory as a function of χ (the size relative to the wavelength) for materials with various specific refractive indexes m, which have no absorption (*i.e.* $Q_{\text{abs}} = 0$, or $k = 0$).
Reproduced from H. C. Van de Hulst, *Light Scattering by Small Particles*, 1981, p. 151 with permission from Dover Publications.

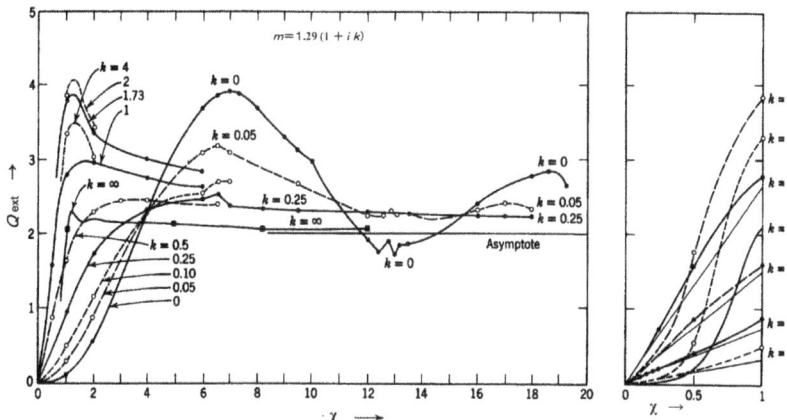

Figure 2.10 The value of Q_{ext} calculated by Mie theory as a function of χ (normalized diameter) for materials that possess various absorption indexes k with specific refractive indexes $m = n^*/n_m = 1.29(1 + ik)$.
Reproduced from H. C. Van de Hulst, *Light Scattering by Small Particles*, 1981, p. 278 with permission from Dover Publications.

dimensionless parameter χ, which describes the particle radius a relative to the wavelength of light λ, *i.e.*, eqn (2.30), is used.

$$x = \frac{2\pi a}{\lambda} = \frac{2\pi a}{\lambda_{vac}/n_m} \tag{2.30}$$

2.4.1 For the Case of Non-absorption

Figure 2.9 shows the extinction factor, Q_{ext}, for the particles with different specific refractive indexes, m, where $Q_{abs} = 0$, and then Q_{ext} is equal to Q_{sca} since $k = 0$ was assumed. When the particle diameter $(2a)$ is equal to $1/\pi$ of the wavelength λ, $\chi = 1$ holds. When the particle becomes larger $(\chi > 1)$, Q_{sca} $(= Q_{ext})$ becomes larger, leading to the strong scattering arising from the interference of the light.

Let's think about the light scattering of TiO_2 particle in aqueous suspension with this figure. Since the refractive index of water n_m is 1.33, the wavelength of the light of 500 nm in air becomes 376 nm $(= 500\,\text{nm}/1.33)$ in water. Since the refractive index n of TiO_2 is about 2.7, the specific refractive index m of TiO_2 in water is close to 2.0 $(= 2.7/1.33)$. The curve for $m = 2$ in Figure 2.9 shows a peak at $\chi = 2.1$. Therefore, using eqn (2.30), $2a = \chi\lambda/\pi = 2.1 \times 376\,\text{nm}/\pi = 250\,\text{nm}$ holds. This means that for a diameter $(= 2a)$ of 250 nm the light of wavelength 500 nm is scattered most strongly for the TiO_2 particle in water.

2.4.2   For the Case of Absorption

When k is not zero, namely absorption takes place, the specific refractive index m becomes a complex number and the feature of the figures becomes largely different. The calculated values of Q_{ext} on changing k are shown in Figure 2.10 for the case in which the real part of the relative refractive index, n/n_m, is 1.29. The relation among the particle size, the wavelength of the light, and the light extinction can be understood roughly from the figure. When the particle size is much smaller than the wavelength (or $\chi < 1$), the following equations can be approximately used,[5]

$$Q_{sca} = \frac{8}{3}x^4 \left| \frac{m^2 - 1}{m^2 + 2} \right|^2 \tag{2.31}$$

$$Q_{abs} = 4x\mathrm{Im}\left(\frac{m^2 - 1}{m^2 + 2} \right) \tag{2.32}$$

where Im means that only the imaginary part is exploited.

2.5   Photoabsorption of Powder Suspension

2.5.1   The Case of Non-agglomeration

Let's calculate the extinction coefficient when the anatase particles of radius a (nm) of $1\,\mathrm{mg\,cm^{-3}}$ concentration are suspended. The crystal density of anatase is $3.894\,\mathrm{g\,cm^{-3}}$ (see Table 1.1). Since the concentration of TiO_2 is $1\,\mathrm{mg\,cm^{-3}}$, the number of particles per $1\,\mathrm{cm^3}$, N, is $1.93 \times 10^{17}/\pi a^3$. Thus, $N\pi a^2$ in eqn (2.28) is estimated to be $1.93 \times 10^3/a$ $(\mathrm{cm^{-1}})$.

For the particle of 24 nm diameter, the radius a is 12 nm and $N\pi a^2$ is calculated to be $160\,\mathrm{cm^{-1}}$. Taking the wavelength of light of 400 nm, with eqn (2.30), $\chi = 0.25$ is obtained. Without particle agglomeration, by using eqn (2.29), (2.31), and (2.32), Q_{ext} can be approximately calculated by using m. The extinction coefficient k at 400 nm (or 3.1 eV) of anatase crystals can be estimated to be 8×10^{-6} for $E//a$ with the value of α in the literature.[7] k for powder is taken to be 1/3 of that as the average. With this the specific refractive index m is calculated to be $m = n*/n_m = 2 + 2 \times 10^{-6}i$. Thus the scattering factor Q_{sca} can be calculated to be 0.0026 from eqn (2.31), and the absorption factor Q_{abs} can be estimated to be 7×10^{-7} from eqn (2.32).

The summation of these values, Q_{ext} ($= Q_{sca} + Q_{abs}$) becomes 0.0026 and the apparent extinction coefficient γ becomes 0.42 cm^{-1}. Therefore, theoretically when anatase particles of diameter 24 nm are suspended in water at the concentration of 1 mg cm^{-3}, light of wavelength 400 nm is extinguished to 1/e ($= 0.37$) by scattering on advancing about 2.4 cm ($= 1/\gamma$), which is called the penetration depth.

2.5.2 The Case of Agglomeration

In practice, in the aqueous suspension of TiO$_2$ the particles are generally agglomerated to form secondary particles of about 1 µm diameter. Let's calculate the apparent absorption when anatase particles of 1 µm diameter are suspended at the concentration of 1 mg cm^{-3}. The ratio of the particle diameter to the wavelength χ becomes 10. As $\chi = 10$ holds, Q_{ext} is assumed to be about 2 from Figure 2.9 for $m = 2$. With this value, the apparent absorption coefficient γ becomes 7.7 cm^{-1} from eqn (2.28). Therefore, the incident depth of the light to the suspension of agglomerates of 1 µm diameter ($1/\gamma$) becomes 1.3 mm, while it is 2.4 cm for the case of non-agglomerated TiO$_2$ of 24 nm diameter. When nano-particles are well dispersed, or non-agglomerated, the suspension becomes transparent.

The application of Mie theory stated above is limited to the spherical single particle and it is not easy to obtain the solution. However, owing to the recent development of numeric calculation, Maxwell's electromagnetic equation could be solved directly[8] by a finite element method without the aid of Mie theory. Since this calculation method is not limited to spherical particles, it can be widely applied.

For the practical photocatalytic systems, only the absorbed light of the irradiated light exerts the reactions while the scattered light is not involved in the reaction. Hence, when the light scattering takes place, the actual absorbance of the light in the particles must be measured. The photoabsorption of the materials which scatter light is measured by a method called the diffuse reflection method. The diffuse reflectance is usually analyzed with Kubelka–Munk theory, which supposes that the reflective samples are thick enough for incident light not to pass through. The absorption coefficient of powders α is described as the following eqn (2.33), using relative reflectance R and scattering coefficient s (cm^{-1}).

$$\frac{\alpha}{s} = \frac{(1 - R)^2}{2R} \tag{2.33}$$

The relative reflectance R is the ratio of the reflectance of the sample powder to that of standard powder with no absorption. As a standard powder, MgO or BaSO$_4$ is employed. However, when the particle size and the refractive index of the sample powder are greatly different, the direction and the extent of the reflection of the light are not the same. Therefore, to average the reflective light in all directions, an integrating sphere is attached to the spectrometer for the measurements. However, it is not easy to obtain the absolute value of α with this method because of the difficulty in the estimation of s.

Relative transmittance, T, can be measured by putting the sample in front of the integrating sphere. Since the fate of the incident light could be absorption (A), transmission (T), and/or reflectance (R), eqn (2.34) holds.

$$A + T + R = 1 \tag{2.34}$$

In the study of photocatalysis, sometimes transmission of the light can be ignored $(T = 0)$. Therefore, the absorbed fraction of the incident light can be calculated by measuring relative reflectance R, using the relation of $A = 1 - R$.

2.6   Retardation of Light Pass (Photonic Crystal)

As described above in Section 2.3 the light interferes with materials of a specific structure to change absorption. The simple example is the interference of light with a thin film as shown in Figure 2.8. This phenomenon is caused by the retardation of the phase velocity in eqn (2.15) because of the formation of a standing wave between two reflective interfaces. For periodic multi-layer film stack, the effect is enhanced, and the material is called a one-dimensional photonic crystal. In nature opal is essentially a natural photonic crystal, which is called structural coloration. Inverse opal structure causes light to remain inside the photocatalytic materials. To enhance the effect of photons, inverse opal photonic crystals showed enormous promising application in optic waveguides, photovoltaic cells, and sensors, as well as photocatalysts.

To prepare three-dimensional photonic crystals of inverse opal, usually polystyrene latex beads were incorporated to form TiO$_2$, for example, and then the beads were removed to make three-dimensionally aligned voids. The photocatalysts in the voids show higher photocatalytic activity owing to the enhancement of light intensity by retarding the phase velocity.[9]

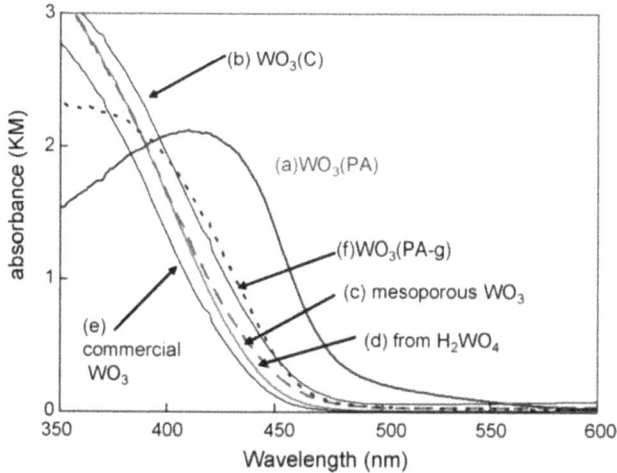

Figure 2.11 UV-Vis absorption spectra of various WO₃ samples. Reprinted from K. Sayama, H. Hayashi, T. Arai, M. Yanagida, T. Gunji and H. Sugihara, Highly active WO₃ semiconductor photocatalyst prepared from amorphous peroxo-tungstic acid for the degradation of various organic compounds, *Appl. Catal., B*, **94**, 150–157. Copyright (2010) with permission from Elsevier.

Another example is the absorption enhancement of WO_3 powders prepared by a special method to form voids at the surface. Figure 2.11 shows the absorption spectra of various WO_3 powders prepared by different methods.[10] The spectrum for (a) WO_3(PA), which is prepared from $[WO_2(O_2)H_2O]$ crystal, shows a remarkable absorption at longer wavelength. When WO_3(PA) was ground, the spectrum changed to the normal feature (f) WO_3(PA-g). The photocatalytic activity is increased by the increase of the absorption.[10]

2.7 Colors of Light and Pigments

The wavelength dependence of the light intensity that is the irradiance spectra for the light sources around us is shown in Figure 2.12. Solar light is basically the spectrum of black body radiation at 6000 °C, which shows the maximum intensity around 500 nm and presents complicated structures due to the absorption by H_2O, O_2, and CO_2 in the air.

The irradiance spectrum of fluorescent light consists of the line spectrum of mercury and the broad emission of fluorescent materials. Black light (BL) is a kind of fluorescent light which is designed to emit

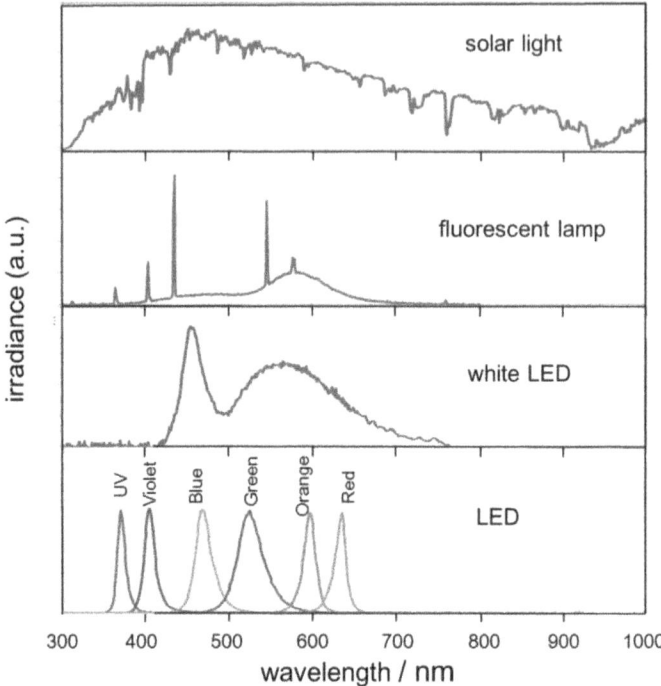

Figure 2.12   Irradiance spectra of several light sources.

only ultraviolet (UV) region of light around 360 nm. Since BL is used as the light source of ISO (the International Organization for Standardization) test for photocatalysts, details of which will be shown in Section 9.2 in Chapter 9. The irradiance of white LED (light emitting diode) consists of the blue LED light and the emission by yellow fluorescent materials as shown in Figure 2.12. Since for LED the energy efficiency of the emission is high and the regulation of the wavelength is easy, it is widely applied to many practical fields, such as traffic signals. For photocatalytic reactions, mercury lamps and xenon lamps were often used in the past. However, recently LED is frequently used as a light source because of the narrow width of the wavelength and low heat generation.

The wavelength of visible light is around 500 nm. This is because we have evolved under the sunlight energy. Therefore, if the surface temperature of the sun were not 6000 °C, different creatures that utilize energy of different wavelengths would have appeared. As our eyesight can respond to the sunlight, we can see light of a broad range of 400–700 nm. The light in this range of the wavelength can be categorized into three ranges with the width of 100 nm, *i.e.*, those around 450, 550, and 650 nm, which we see as blue, green, and red colors, respectively. These colors are called the three primary colors of light and they are abbreviated as B, G, and R, respectively. The white color W is the summation of these three kinds of color and expressed as follows.

$$R + G + B = W$$

On combining two of the three primary colors, Yellow (Y), Magenta (M), and Cyan (C) are obtained. These colors are also obtained by subtracting the lights of B, G, and R from the white color W, as expressed as follows.

$$R + G = Y = W - B$$

$$R + B = M = W - G$$

$$G + B = C = W - R$$

Y, M, and C are the primary colors of the color materials (or ink) used for printers. These absorption spectra are shown schematically in Figure 2.13. Namely, Y ink absorbs just B color, and so on.

The colors that appear on the prints by absorbing the colors are reflected from white paper with the color materials. For instance, the color of R can be created by putting the color materials of Y and M on W. Namely,

$$Y + M - W = R$$

$$Y + C - W = G$$

$$M + C - W = B$$

The semiconductors used for photocatalysts present colors when the transition energy between the bands corresponds to the visible light range, *i.e.*, 1.8–3.1 eV. With the elongation of the wavelength of

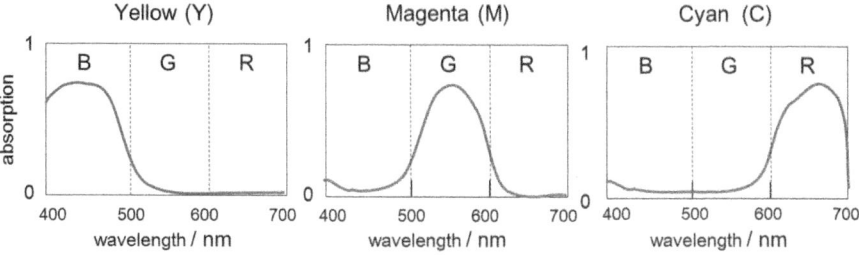

Figure 2.13 Schematic absorption spectra of three kinds of dyes (Y, M, and C) of printer ink, showing selective absorption at the region of blue (B), green (G), and red (R).

the absorption edge, by absorbing the light in the range of B, B + G, and B + G + R successively, the colors change to yellow, orange, red, brown, and black. Therefore the other colors are not presented.

References

1. K. Vos and H. J. Krusemeyer, *J. Phys. C: Solid State Phys.*, 1977, **10**, 3893.
2. H. Tang *et al.*, *Solid State Commun.*, 1994, **92**, 267.
3. A. R. Forouhi and I. Bloomer, *Phys. Rev. B: Condens. Matter Mater. Phys.*, 1986, **34**, 7018.
4. S. Sakka, *Science of Sol-gel* (in Japanese), Agune-Shofu-Sha, Tokyo, 1988.
5. H. C. van de Hulst, *Light Scattering by Small Particles*, Dover Publ. Inc., New York, 1981.
6. M. Born and E. Wolf, *Principles of Optics*, Pergamon Press, Oxford, 1959.
7. T. Sekiyta *et al.*, *J. Phys. Chem. Solids*, 2000, **61**, 1237.
8. L. E. McNeil, A. R. Hanuska and R. H. French, *Appl. Opt.*, 2001, **40**, 3726.
9. Z. Geng, Y. Lu *et al.*, *J. Alloys Compd.*, 2015, **644**, 734.
10. K. Sayama *et al.*, *Appl. Catal., B*, 2010, **94**, 150.

3 Principles of Semiconductors

3.1 The Electronic Energy of the Semiconductor

All substances consist of atoms. Recall that an atom consists of a positively charged nucleus and some electrons, the number of which is equal to that of the positive charge or the atomic number. The basic formation of solid materials takes a crystalline structure in which atoms are lined up three dimensionally.

To understand the electronic energy in a solid, let's start with two atoms, A and B, which accompany one valence electron of the s-orbital for each atom. As shown in Figure 3.1, when the distance between the atoms A and B becomes smaller to form a molecule A–B, electrons of the atoms interact with each other to generate two energy states that are bonding and anti-bonding states. For the bonding orbital with low energy, electrons are distributed among atoms, while for the anti-bonding orbitals there is a node of the orbital function where the distribution of electrons becomes zero. Electron distribution and the coefficient of each atomic orbital are presented on the right side in Figure 3.1.

Let's consider titanium dioxide (TiO_2). The atomic number of titanium, Ti, is 22. The electron configuration in atomic orbitals is $(1s)^2(2s)^2(2p)^6(3s)^2(3p)^6(4s)^2(3d)^2$. On the other hand, the atomic number of the oxygen atom is 8 and the electron configuration is $(1s)^2(2s)^2(2p)^4$. The electron configuration of ionic Ti^{4+} is $(1s)^2(2s)^2(2p)^6(3s)^2(3p)^6$ and that of O^{2-} is $(1s)^2(2s)^2(2p)^6$. The energy in atomic states is roughly shown on the left side in Figure 3.2. When atoms are ionized, the energy level of O $2p_z$ increases as compared to

Introduction to Photocatalysis: From Basic Science to Applications
By Yoshio Nosaka and Atsuko Nosaka
© Yoshio Nosaka and Atsuko Nosaka, 2016
Published by the Royal Society of Chemistry, www.rsc.org

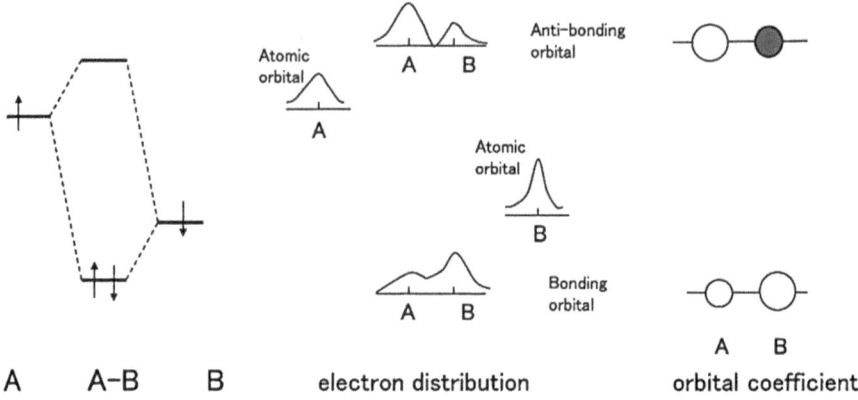

Figure 3.1 By encountering an atom A and an atom B with s-orbitals, two electronic levels are formed, which are expressed by the electron distribution and the coefficient of atomic orbitals.

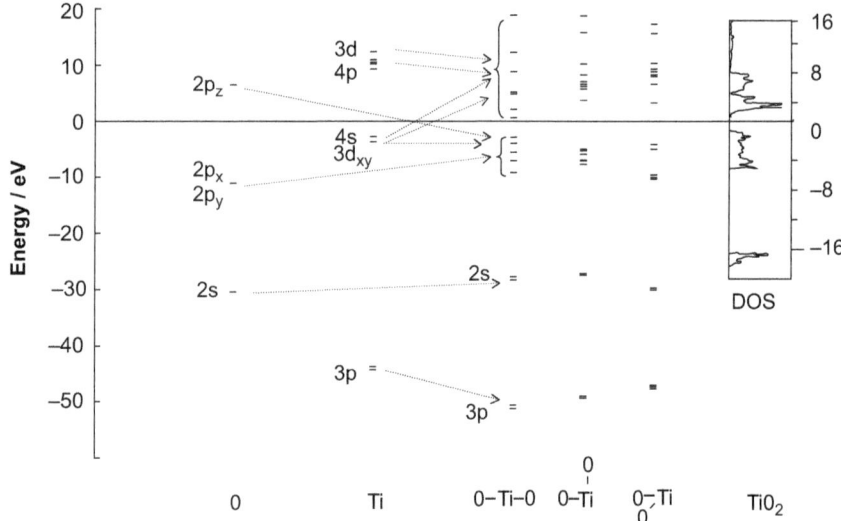

Figure 3.2 Electronic energy of oxygen and titanium atoms, and energy levels of three virtual TiO_2 molecules of different structures[1] showing correlation with DOS of anatase TiO_2 crystal.
Reprinted (figure) with permission from R. Asahi *et al.*, *Phys. Rev. B: Condens. Matter Mater. Phys.*, **61**, 7459. Copyright 2000 by the American Physical Society.

$2p_x$ and $2p_y$, while those of Ti 4s and $3d_{xy}$ increase as compared to the other 3d-orbitals.

The energy level of the virtual molecule in which one Ti atom binds with two oxygen atoms is calculated by an *ab-initio* molecular orbital method as shown in Figure 3.2.[1] By calculating of the electronic energy by changing the relative position of oxygens bound to Ti, it was revealed that the relative position of O atoms barely affects the bonding since the energies of O 2s- and Ti 3p-orbitals are quite low. Furthermore, for the bonding orbital whose energy is less than zero the contribution of O 2p-orbital is large, whereas for the anti-bonding orbital the contribution of 4s- or 3d-orbitals of Ti is large. Because such a bonding exists to a large extent for solids, the energy levels can be described only by the number of the levels in a certain energy interval, or density of states (DOS). The bundle of the energy levels is called an energy band. The DOS of anatase TiO_2 calculated by FLAPW (full-potential linearized augmented plane-wave) method[2] is shown on the right side of Figure 3.2. Electrons usually reside in the levels whose energy is lower than zero and do not exist on the levels whose energy is higher than zero.

Since the electronic states of solids are comprised of the combinations of a large number of orbitals, the actual state can be grasped only by DOS. Then, the way of the combination of electrons among the atoms is determined by the periodic structure of atomic orbitals in the different directions in the solid. The way of the combination is distinguished by the wave number (or wave vector) k. In other words, electrons in solids distribute through the solids owing to the electronic interactions among the atoms, while some electrons are captured by the individual atomic nuclei because the atomic nuclei possess large positive charge. Then, a wave function called the Bloch function, which describes both the part restrained by atoms and that distributing in the solid, is used.

On considering the relationship of the band energy with k, it would be easier for understanding to begin with one-dimensionally lined atoms.[3] In Figure 3.3 the extent of the contribution of the s-orbital of each atom is indicated by the size of the circles when atoms are in a line at the regular intervals of b. The dotted wave shows the envelope function of the Bloch function, which describes whole extension in the direction of the line. In general, free electrons in solids are described as sine wave of the envelope function and the momentum becomes smaller with an increase of wavelength λ. The wave number k $(=2\pi/\lambda)$ is correlated with the number of nodes, that is the points where the envelope function changes the sign. For free electrons with infinite wavelength of the envelope function, $k=0$ holds and all the s-orbitals overlap to show uniformly vast distribution of electrons among the atoms. When the neighboring atoms are all of anti-bonding characteristics, $k=\pi/b$ holds and atoms generate the anti-bond repeatedly. This means that, for the s band consisting of the row of s-orbitals, the energy is the lowest at $k=0$ and the highest

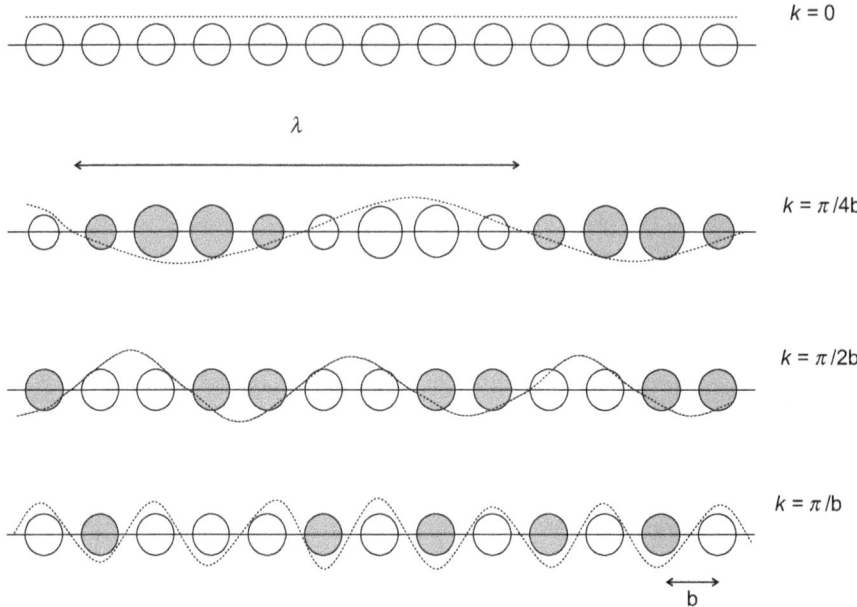

$k = 0$

λ

$k = \pi/4b$

$k = \pi/2b$

$k = \pi/b$

b

Figure 3.3   Modes of wave functions for the row of 1s atomic orbitals represented by circles corresponding to the coefficient at each atom.

at $k = \pi/b$. The situation is illustrated at the bottom of Figure 3.4C with corresponding schematic DOS.

Figure 3.4A and B show the interaction of p-orbitals when atoms are in line. The features in which p-orbitals are oriented parallel to the atomic row ($p_{//}$) are shown in Figure 3.4A. Since all the neighboring atoms become anti-bonding at $k = 0$, the energy becomes the highest (a), while at $k = \pi/b$ the σ bonds are formed among all the atoms and the energy becomes the lowest (b). On the other hand, as shown in Figure 3.4B, for $p_\perp$-orbitals which are oriented vertically to the atomic row, at $k = 0$ every neighboring p-orbital forms π bonds (c), while at $k = \pi/b$, the anti-bonding π-orbitals (d) are formed, showing the k dependency similar to the s band. It is called an energy band because each group of the energy levels possesses a certain energy width. Thus, when atoms having electrons in the s- and p-orbitals make a one-dimensional row, the E–k diagram and DOS would be depicted as shown in Figure 3.4C.

To be familiar with the band theory, let's extend the one-dimensional row of atomic orbitals to a two-dimensional array. There are two directions, x and y, to describe the k vector. Figure 3.5A shows four combinations of (k_x, k_y) with 0 and π/b. In Figure 3.5B, the k dependence of the energy is schematically illustrated.

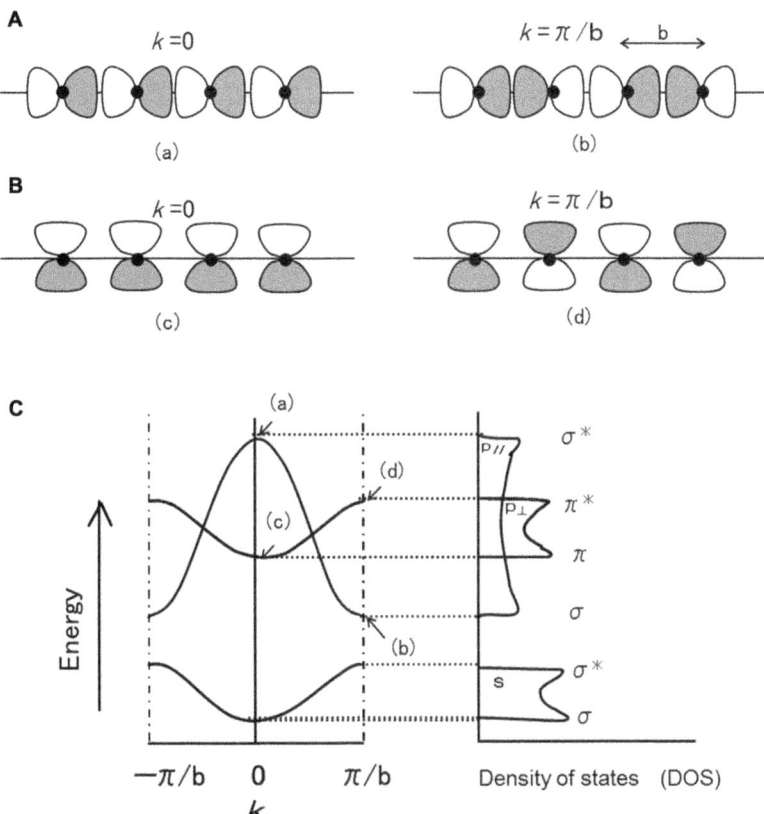

Figure 3.4 Schematic illustration for p atomic orbitals directed (A) parallel and (B) perpendicular to the row of atoms with coefficients of the same sign ($k=0$) and the alternative sign ($k=\pi/b$). (C) One-dimensional band formation (E–k diagram) for s and p atomic orbitals where the points (a)–(d) correspond to those in (A) and (B).

Since in physics the wavenumber k corresponds to the momentum of the particle, the energy E of the particle can be calculated from the momentum k with the apparent mass of the particle or effective mass m^* as expressed in eqn (3.1).

$$E(k) = \pm \frac{\hbar^2 k^2}{2m^*} \tag{3.1}$$

Here, $\hbar k$ represents the momentum p of the particle, which is obtained by rearranging the well-known de Broglie equation $p = h/\lambda$ with $k = 2\pi/\lambda$ and $\hbar = h/2\pi$. The m^* is usually expressed as a relative value to electron rest mass m_0 ($= 9.11 \times 10^{-31}$ kg). For simplicity, m_0 is usually omitted in equations, but for the numerical calculation the

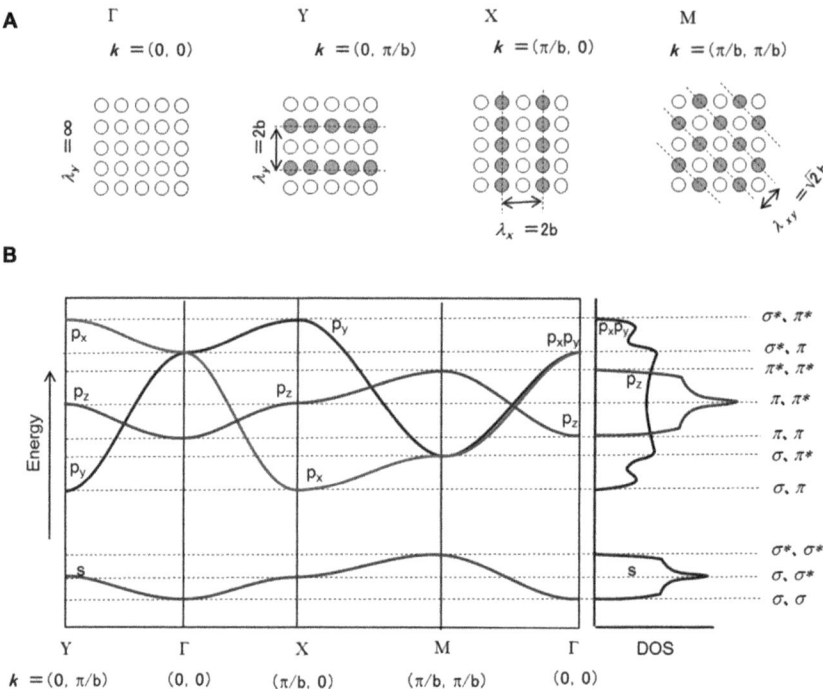

Figure 3.5   (A) Two-dimensional array of 1s atomic orbitals at four sets of values of (k_x, k_y). Closed and open circles indicate the sign of the coefficient at each atomic orbital, respectively. (B) Schematic $E–k$ plots for two-dimensional array in x–y plane for s, p_x, p_y, and p_z atomic orbitals.

m^* should be multiplied by m_0. As shown in Figure 3.5B, the curve of *E versus k* at around $k=0$ can be expressed by eqn (3.1). Therefore, the smaller the effective mass, the larger is the curvature in the *E–k* figure. Namely, the energy changes largely with a small change of the momentum. Or, a less heavy m^* gets a lower momentum $\hbar k$ with a given energy.

In the three dimensional solids, k is described as a vector. Therefore, the effective mass is defined as a reciprocal of the second derivative of energy and expressed by a tensor in a strict sense. Thus, the elements of the tensor can be expressed as eqn (3.2), where i and j represent the direction of the k vectors or the direction of the crystal axes in many cases.

$$m_{ij}^* = \frac{\hbar^2}{2\left(\dfrac{\partial^2 E}{\partial k_i k_j}\right)} \tag{3.2}$$

Experimentally, the effective mass is measured by a spectro-electrochemical method or by the Hall effect, that is the change of the conductivity in a magnetic field.

Now we could understand that the electron energy in a semiconductor can be distinguished by using the wave vector k, which is described by the direction of atomic line and the kind of atomic orbitals.

Electrons in a solid occupy successively the energy levels from the lower to the upper. For the case of solids, the band occupied with electrons is called the valence band (VB), while the lower level barely occupied with electrons is called the conduction band (CB). The energy gap between these two bands is called the bandgap energy (E_g). Electric properties of materials can be classified by the position of the Fermi level (E_F) against the bands. The Fermi level is the energy level below which electrons occupy the levels but above which they are vacant. Naturally, the Fermi level is diffused by thermal energy of $k_B T$ with the Boltzmann constant $k_B = 1.38 \times 10^{-23}$ JK^{-1}, which is calculated to be 0.026 eV at 25 °C.

Figure 3.6 shows an insulator, a conductor, and n-type and p-type semiconductors. For instance when electrons fill half of one band, they can move readily in the material. Namely, for a good conductor, the Fermi level is located in a band. On the other hand, when the

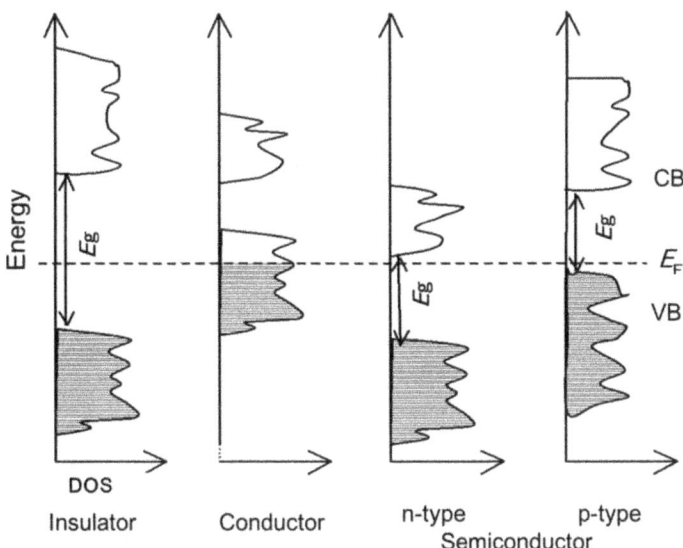

Figure 3.6 Relationship between the electrical specifications of materials and the position of energy bands relative to the Fermi level (E_F).

Fermi level is located in the bandgap, the lower band is completely occupied by electrons but the upper band is not occupied by electrons at all. In this situation, electrons cannot move easily and then the material becomes an insulator. For semiconductors, since the bandgap is not large, two kinds of electric currents can be generated, *i.e.*, electrons (negative charge) flow through the conduction band above the Fermi level, or positive charges (positive holes) flow through the valence band under the Fermi level. The former is called an n-type semiconductor and the latter a p-type semiconductor. Titanium dioxide is an n-type semiconductor, because mobile electrons are generated when the amount of oxygen in TiO_2 becomes slightly less than the stoichiometric ratio. Since the extent of dispersion of the electron energy at the Fermi level obeys a Boltzmann distribution, the conductivity of the semiconductor increases with the increase of the temperature, which is opposite to the case of common resistors.

3.2   Absorption of Light by Semiconductors

In general, light absorption by materials is a phenomenon that occurs under an oscillated electromagnetic field, in which several types of energy transfer proceed. Figure 3.7 shows a representative absorption spectrum of a semiconductor in the wide range of wavelengths.

In the lowest energy region, free carrier absorption is observed, which can be explained by a classical theory of electrodynamics.

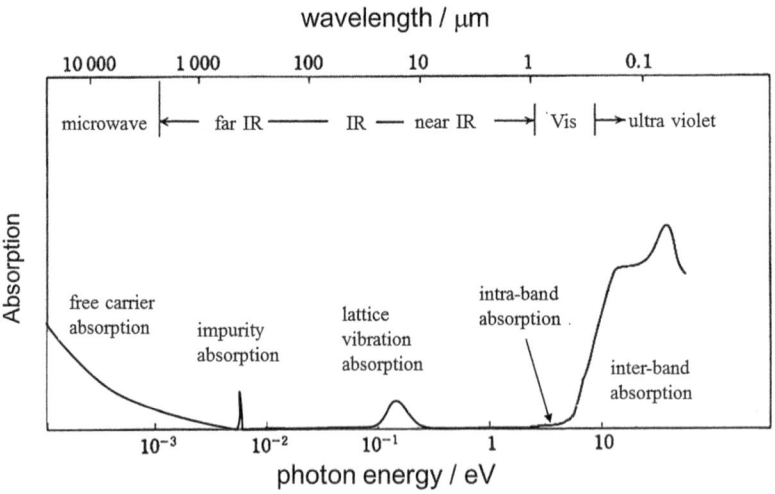

Figure 3.7   Representative absorption spectrum of a semiconductor.

The principle of the absorption is that the electric field of the light causes the motion of free carriers based on the Drude model. Firstly the complex permittivity ε^* is described by eqn (3.3),[4]

$$\varepsilon^*(\omega) = 1 - \frac{\omega_p^2}{\omega\left(\omega + \dfrac{i}{\tau}\right)} \tag{3.3}$$

where τ is the lifetime of the free carrier, and plasma frequency ω_p is defined by eqn (3.4)

$$\omega_p^2 = \frac{e^2 N_0}{\varepsilon_0 m^*} \tag{3.4}$$

with the carrier density N_0, elementary charge e, pemittivity of vacuum ε_0 and the effective mass m^*. By exploiting eqn (2.22) in Chapter 2, the absorption coefficient α can be obtained from the imaginary part ε_2 of eqn (3.3), which is in turn expressed as eqn (3.5), by assuming the mobility of the carrier as $\mu = e\tau/m^*$ with $\tau \gg 1/\omega$, and substituting ω by $\lambda = 2\pi c/\omega$.

$$\alpha = \frac{\omega \varepsilon_2}{cn} = \frac{e^3 \lambda^2 N_0}{4\pi^2 c^3 \varepsilon_0 \, m^{*2} n\mu} \tag{3.5}$$

Thus, the free carrier absorption increases with the increase of wavelength λ.

In far IR regions, the absorption of impurities such as donors and acceptors can be observed. Furthermore, several types of lattice vibration cause IR absorption, which is called an "optical phonon". Similarly to the IR absorption, Raman spectroscopy in these wavelength regions can be utilized to analyze crystal structures.

The main absorption, that is the most important absorption in photocatalysis, is the inter-band absorption, arising from an excitation of electrons in the valence band to the conduction band. Simultaneously, holes in the valence band are generated. The electrons and holes generated in the photoabsorption are called photoinduced carriers together. For atoms and molecules, electron excitation takes place between specific energy levels of narrow width to provide relatively sharp spectra. However, for semiconductors the energy levels of electrons are remarkably wide as shown in Figure 3.6, resulting in photons with energy more than the bandgap to be absorbed in the remarkably wide range in the spectrum. Titanium oxide is usually transparent in the visible light region because the bandgap energy is in the range of ultraviolet wavelength. Powders of TiO_2 look white, because they scatter all the photons of visible wavelength region. Scatter means that light is reflected to random directions due to the inhomogeneity in the surface configuration. When the bandgap becomes narrower than 1.7 eV, the semiconductor absorbs photons of all visible wavelengths, then it colors black. Therefore, along with widening the bandgap from 1.7 to 2.5 eV, the powder of a semiconductor provides a series of colors, *i.e.*, brown, red, orange, and yellow as was described in Section 2.7.

The magnitude of the photoinduced electron transition is determined by the dipole moment caused by the transition at the crystal unit cell. As schematically shown in Figure 3.8, the transition moment from the $k=0$ state of the s band to the $k=0$ state of the $p_\perp$ band is large. On the other hand, the transition moment to the $k=\pi/b$ state of the $p_\perp$ band becomes alternative for each atom, resulting in no net transition. However, when the lower level is the $k=\pi/b$ state of the s band, the excitation to the $k=\pi/b$ state of the $p_\perp$ band is allowed. Thus, the transition is enabled for the same k wave vector. Since the photoabsorption of the semiconductor solid is the sum of the transition dipole moments of all of the unit cells, the absorbance of the materials per weight does not change. In other words, when the particle size becomes larger, the absorption increases only due to the increase of the number of the crystal unit cells. For organic molecules, in general, the absorbance does not depend on the molecular size, but on the number of chromophores in the molecule.

Inter-band absorption can be classified into two types by the E–k diagram. As shown in Figure 3.9A, when the energy of the ground state gives maximum (top of the valence band) at the wave vector k at which the energy of the excited state gives minimum (bottom of the conduction band), direct electron transfer occurs for the same k. Such a semiconductor is called a direct bandgap semiconductor.

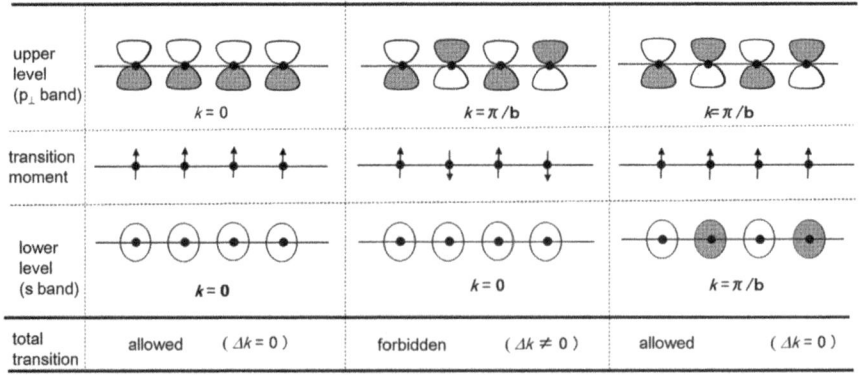

Figure 3.8 Schematic illustrations showing optical transitions between band levels.

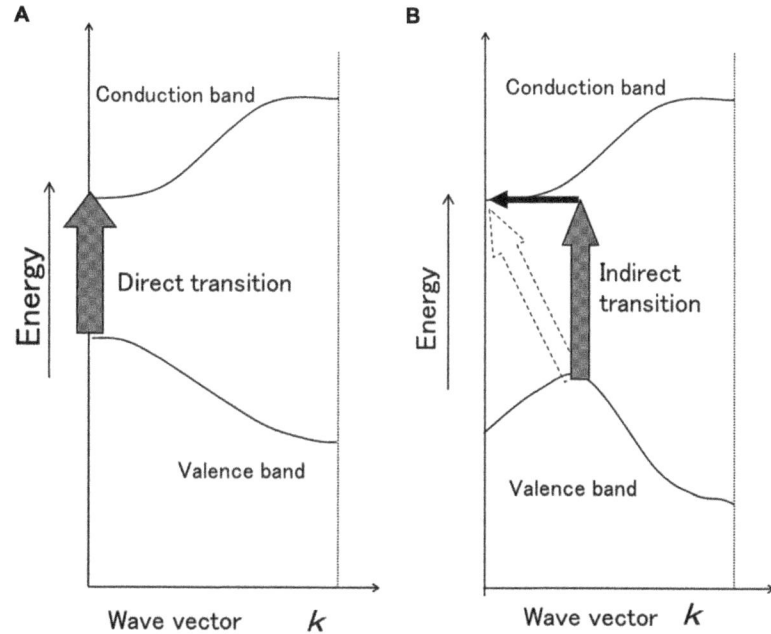

Figure 3.9 The absorption processes of light by (A) direct bandgap semi-conductors, and (B) indirect bandgap semiconductors.

However, in general, the maximum of the valence band is not always the minimum of the conduction band at any k. Namely, as shown in Figure 3.9B, the electronic excitation to the conduction band minimum cannot occur, because k should be changed ($\Delta k \neq 0$). In this case, the light absorption takes place actually along with the vibrational excitation of the crystal lattice ($\Delta k = 0$), *i.e.*, the absorption process involves the excitation of phonon to change the wave vector k. Thus, it is called an indirect bandgap semiconductor. Therefore, for an indirect bandgap semiconductor, the absorption edge of the absorption spectrum becomes gentle as compared to that of a direct bandgap semiconductor.

For the absorption spectra of TiO_2, as shown in Figure 2.6A, the absorption edge for anatase polymorph is gentle; therefore, anatase seems to be an indirect bandgap semiconductor. The details of the band energy diagram (E–k diagram) calculated for rutile and anatase TiO_2 polymorphs are shown in Figure 3.10 A and B, respectively.[5] For rutile the direct transition at $k = \Gamma$ and for anatase the indirect transition from Z to Γ are expected. The symbols in the abscissa correspond to the wave vectors by which the direction of the wave vectors of the atomic arrays in the crystal is discriminated.

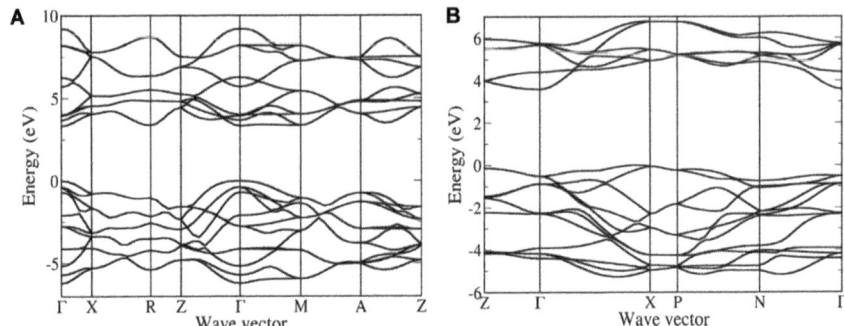

Figure 3.10 Band energy diagram (*E–k*) of (A) rutile and (B) anatase TiO₂ calculated by the PAW method.
Reprinted with permission from P. Deak *et al.*, *J. Phys. Chem. C*, 2011, **115**, 3443.[5] Copyright 2011 American Chemical Society.

For the direct bandgap transition, according to eqn (3.1), energy levels of the upper edge of the valence band, E_v, and the bottom of the conduction band, E_c, at around $k = 0$ are expressed as functions of the wave number k as follows,

$$E_v(k) = -\frac{\hbar^2 k^2}{2m_h{}^*} \tag{3.6}$$

$$E_c(k) = E_g + \frac{\hbar^2 k^2}{2m_e{}^*} \tag{3.7}$$

where $m_h{}^*$ and $m_e{}^*$ are the effective masses of the valence band hole and the conduction band electron, respectively, and E_g is the bandgap energy. With these equations, the energy dependency of the state density can be calculated. That is, the relation of the energy difference with the transition probability, which is the wavelength dependency of the absorbance, can be calculated. In general, at the absorption edge of the transition between bands, the absorption coefficient α follows the equation below,

$$\alpha \propto (h\nu - E_g)^n / h\nu \tag{3.8}$$

where $h\nu$ is the energy of photons. For the allowed direct transition from E_v of eqn (3.6) to E_c of eqn (3.7), n in eqn (3.8) should be 1/2. For the indirect transition, $n = 2$. Thus, E_g can be obtained by plotting $(h\nu\,\alpha)^{1/n}$ against $h\nu$ and extrapolating the line to the axis.

3.3 The Potential Photocatalysts Besides TiO$_2$

A number of semiconductors have been examined as potential photocatalysts. Figure 3.11 shows the bottom edge of the conduction band, the upper edge of the valence band, and their difference, *i.e.*, the bandgap energy E_g, for popular semiconductors and some of the semiconductors which were examined as photocatalysts. Since the photocatalytic decomposition of water gathers considerable attention as will be described later, the oxidation and reduction potentials of water are depicted with dotted lines in the figure. For comprehensive data, one can refer to a review report.[6]

Figure 3.12 shows the relationship between the flat-band potential U_{fb} and the bandgap energy E_g of metal oxides.[7] U_{fb} is measured electrochemically and, for metal oxides n-type semiconductors, it is just below the conduction band minimum as will be described in the next chapter. As shown in Figure 3.12, the bandgap energy has a linear relationship with U_{fb} and many metal oxides are located on the line of 45°. This means that the energy of the valence band maximum is not different largely among the different materials. As stated above, the valence band of metal oxides mainly consists of the 2p atomic orbital of oxygen, supporting the above observations. However, a

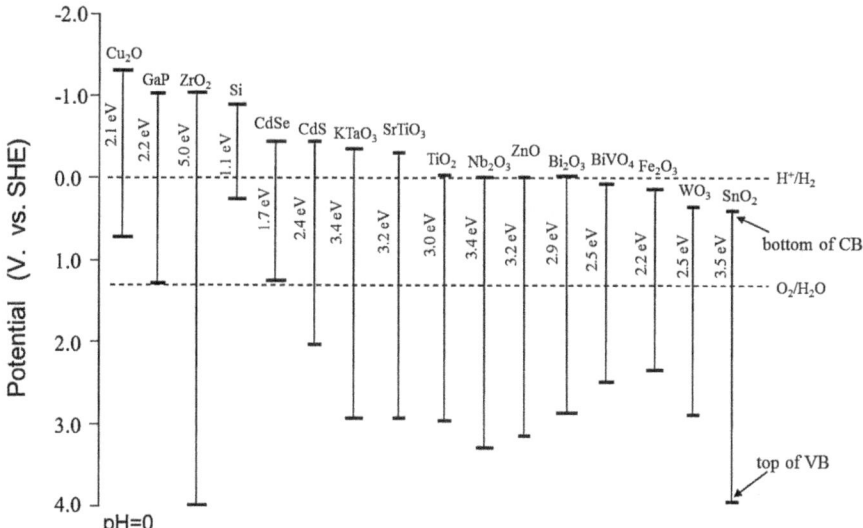

Figure 3.11 Energy levels of the bottom edge of the conduction band (CB) and the upper edge of the valence band (VB) of semiconductors with bandgap energy.

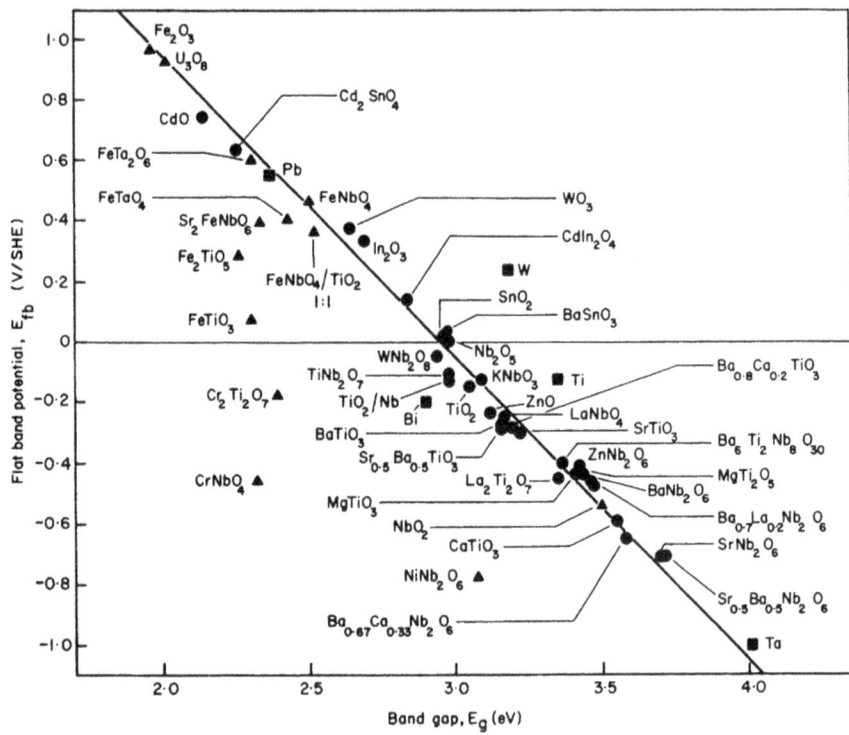

Figure 3.12   Measured flat-band potentials and effective bandgaps in semiconducting oxides. ●, oxides without partly filled d-levels; ▲,
oxides with partly filled d-levels; and ■, oxides formed by
anode oxidation on metals.
Reprinted from D. E. Scaife, Oxide semiconductors in photoelectrochemical conversion of solar energy, *Solar Energy*, **25**,
41–54. Copyright (1980) with permission from Elsevier.

recent study of the band energy for individual metal oxides shown in
Figure 3.11 does not always hold this explanation.

Besides TiO_2, various metal oxide semiconductors have been tested
as photocatalysts. Plotted in Figure 3.12 are the metal oxides with
vacant d-levels. Partly filled d-electrons are rare, since the partly filled
d-electrons are suggested to generate a mid gap level, which may
shorten the lifetime of the photoexcited electron-hole pairs. There-
fore, the metal oxides of d^0 configuration (Ti^{4+}, Zr^{4+}, Nb^{5+}, Ta^{5+}, and
W^{6+}) which do not have mid gap level have actually been employed in
photocatalytic researches.

On the other hand, fully filled d-electrons may not affect the stability of the electron-hole pairs. Therefore, the various typical metal
oxides with d^{10} configuration (Ga^{3+}, Ge^{4+}, In^{3+}, Sn^{4+}, and Sb^{5+})

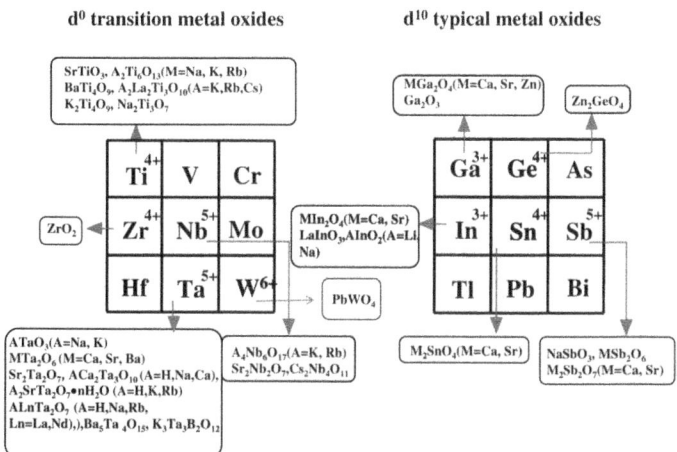

Figure 3.13 Groups of photocatalysts distinguished by d⁰ and d¹⁰ configuration.
Reprinted from ref. 8 with permission from the Royal Society of Chemistry.

together with $d^{10}s^2$ (Pb^{2+}) metal oxides were examined anew as photocatalysts (Figure 3.13).[8] Thus, zinc gallate ($ZnGa_2O_4$), various indates (MIn_2O_4 (M = Ca, Sr), $NaInO_2$, $LaInO_3$), strontium stannate (Sr_2SnO_4), and various antimonates ($M_2Sb_2O_7$ (M = Ca, Sr), $CaSb_2O_6$, $NaSbO_3$) have been found to be photocatalytically active.[9]

In the case of d^{10} metal oxides, the conduction band consists of hybridized sp-orbitals and has a large energy dispersion which causes small effective mass, or large mobility, for photoexcited electrons. Since the bandgap of these d^{10} semiconductors is usually wide, the practical applications would be limited. However, the extension of the selections of metal ions to be used as photocatalysts must open the way to choose appropriate metal ions in exploring new photocatalysts such as nitride.

There exist hundreds of elements on earth, and how they can contribute to the construction of heterogeneous photocatalytic materials is shown in Figure 3.14.[10] The elements are classified into four groups: (i) to construct crystal structure and energy structure, (ii) to construct crystal structure but not energy structure, (iii) to form impurity levels as dopants, and (iv) to be used as co-catalysts.

The valence bands of metal sulfide and nitride photocatalysts are usually composed of S 3p and N 2p-orbitals, respectively. Orbitals of Cu 3d in Cu^+, Ag 4d in Ag^+, Pb 6s in Pb^{2+}, Bi 6s in Bi^{3+}, and Sn 5s in Sn^{2+} can also form valence bands in some metal oxide and sulfide photocatalysts. Alkali, alkaline earth, and some lanthanide ions do

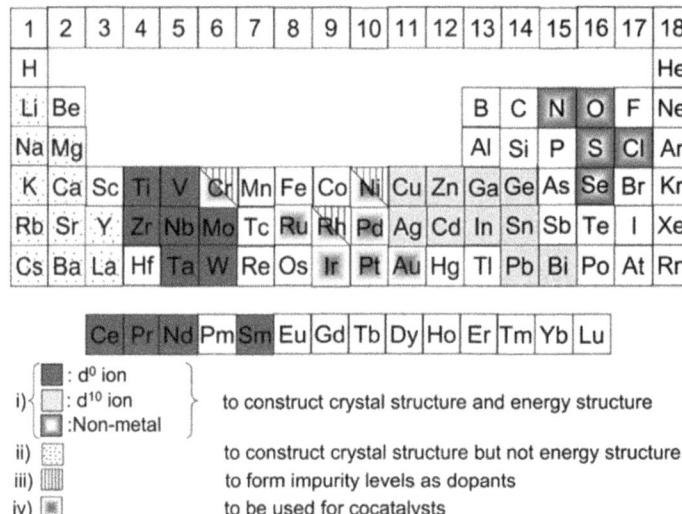

Figure 3.14 Elements constructing heterogeneous photocatalysts. Reprinted from ref. 10 with permission from the Royal Society of Chemistry.

not directly contribute to the band formation and simply construct the crystal structure as A-site cations in perovskite compounds. Some transition metal cations with partially filled d-orbitals such as Fe^{3+}, Cr^{3+}, Ni^{2+}, and Rh^{3+} form some impurity levels in the bandgaps when they are doped or substituted for native metal cations as will be described in Chapter 7. Although they often work as recombination centers between photogenerated electrons and holes, they sometimes play an important role in visible light response. Some transition metals and the oxides such as noble metals (Pt, Rh, and Au), Co_3O_4, NiO, and RuO_2 function as co-catalysts for H_2 evolution.

In the literature, the popular photocatalyst used in documents cited in SciFinder of the American Chemical Society was TiO_2. It appears that 45% of the documents concerned with photocatalysis dealt with TiO_2. The frequency of metal oxides is as follows: $ZnO(10\%)$, $Fe_2O_3(2.0\%)$, $WO_3(1.7\%)$, $Cu_2O(1.7\%)$, $CuO(1.2\%)$, $BiVO_4(1.2\%)$, $Bi_2WO_6(1.2\%)$, $CeO_2(0.8\%)$, and $SrTiO_3(0.8\%)$. Naturally the crystal structure and the electronic structure as well become complicated.[9] For example the crystal structure of $BiVO_4$ and $RbLaNb_2O_7$ are shown in Figure 3.15.

As will be described in Chapter 10, the important application of photocatalysis is to utilize solar light. Therefore, sulfides and the other non-oxide semiconductors are also employed as photocatalysts. They are $CdS(3.8\%)$, $CdSe(3.8\%)$, $ZnS(1.6\%)$, $ZnSe(1.6\%)$,

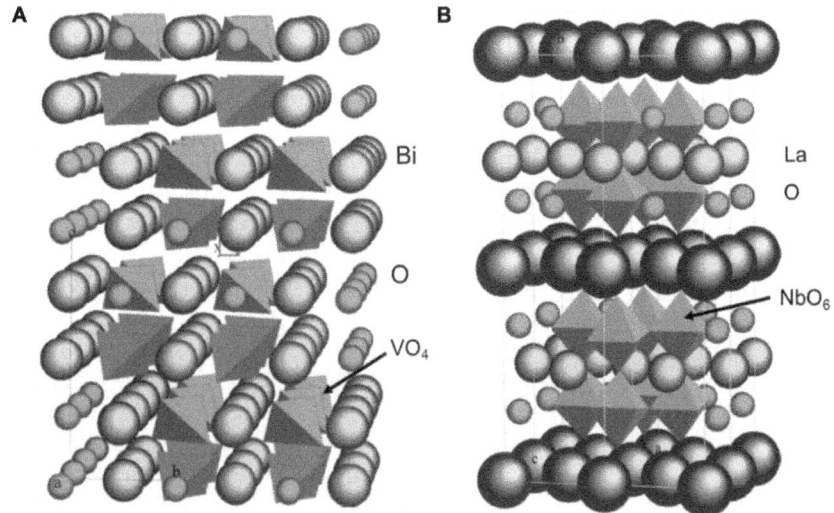

Figure 3.15 Crystal structures of (A) $BiVO_4$ scheelite structure and (B) $RbLaNb_2O_7$ Dion–Jacobson phase.
Reprinted with permission from C. Kormann et al., *J. Phys. Chem.*, 1988, **92**, 5196.[9] Copyright 2012 American Chemical Society.

$Ag_3PO_4(1.1\%)$, α-$C_3N_4(2.7\%)$, and $BiOCl(1.0\%)$. Since the corrosion or dissolution of these solid materials becomes a problem, there is limitation in using solely these non-oxide semiconductors.

3.4 Quantum Size Effect

When the size of a semiconductor becomes extremely small, the energy levels in the band become discrete and they shift due to the confinement of the carriers in the particle, which is called the quantum size effect. The size of the exciton, namely the spread of the conduction band electron or valence band hole generated in the semiconductor by photoabsorption, which is called the Bohr radius, can be calculated by $(\varepsilon/m^*)\times 52.9$ pm. Here, m^* is the effective mass (the unit is electron rest mass m_0) and ε is usually a static dielectric constant. The exciton size of anatase TiO_2 $(\varepsilon = 12 \pm 2, m^* = 0.8 \pm 0.2)^{12}$ is 0.8 nm while that of rutile $(\varepsilon = 113, m^* = 20)$ is quite small at 0.3 nm. When the size of the semiconductor becomes nearly as small as that of the exciton size, the change in the energy structure (quantum size effect) takes place. Namely, owing to the confinement effect, the energy levels shift by the additional kinetic energy V, which can be expressed by eqn (3.9) for the lowest state,

$$V = \frac{h^2}{8m^*a^2} \qquad (3.9)$$

where h is Planck's constant and a is a particle radius. Because this additional energy is caused for electrons in the conduction band and holes in the valence band, the lowest excitation energy of the semiconductor nanoparticles is enlarged from the bandgap energy E_g of the bulk crystal by the quantum size effect ΔE,

$$\Delta E = V_e + V_h + E_{Coul} \qquad (3.10)$$

where V_e and V_h are the energy shift of the conduction band electron and the valence band hole, respectively, and E_{Coul} represents the Coulomb energy of the electron-hole pair in the particle. In the effective mass approximation of eqn (3.9), the relationship of energy shift ΔE (in units of eV) with the particle radius a (nm) is calculated by eqn (3.11),[11]

$$\Delta E = \frac{0.376}{\mu^* a^2} - \frac{2.59}{\varepsilon\, a} \qquad (3.11)$$

where μ^* is the reduced effective mass, which is calculated by $1/\mu^* = 1/m_h^* + 1/m_e^*$ with the effective mass of hole and electron, m_h^* and m_e^*, respectively.

The observed lowest excitation energies of TiO_2 were compared in Figure 3.16A with those estimated by theoretical calculations which took account of the size confinement of a semiconductor. Because the conduction band of TiO_2 comprises d-orbitals, the effective mass m_e^* is expected to be larger than m_h^*. Namely, $m_e^* > 10$.[12] As shown in the figure, for anatase ΔE becomes 0.063 eV with the particle size of 4 nm (radius $a = 2$ nm), while for rutile the energy shift is not observed. Thus due to a relatively large effective mass, the quantum size effect does not appear for TiO_2 because the minimum size of TiO_2 crystallite to be prepared was about 6 nm.

When the effective mass of carriers in a semiconductor is small, e.g., $m^* \sim 0.2$ for ZnO, the shift of the energy band level may become large. In this case, V is overestimated in eqn (3.9), because in the derivation of the equation, the infinite potential has been assumed outside the particle. Although it is not easy to formulate the energy shift V in a finite potential V_0, it can be expressed by eqn (3.12) with

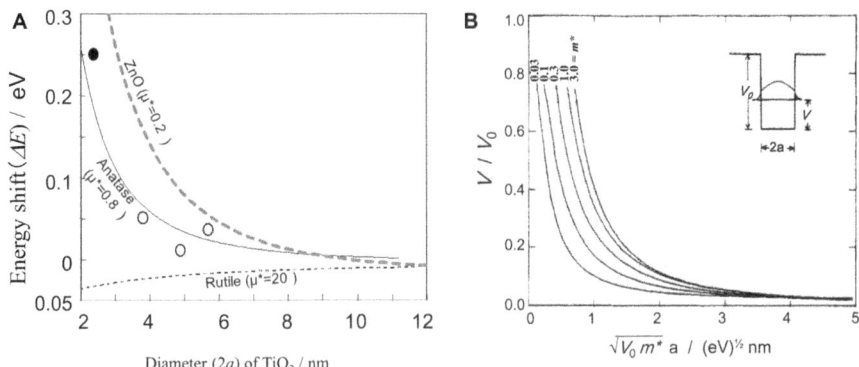

Figure 3.16 (A) Shift of the lowest excitation energy of anatase (solid line) and rutile (dotted line) TiO_2 calculated by eqn (3.10) and (3.12), and experimental data[13,14] for anatase TiO_2. Dotted curve is for ZnO nanoparticle. (B) Energy shift and wave function for 1s level of semiconductor particles calculated by a finite depth square well potential model. Relative energy shift V is characterized by the potential depth V_0, effective mass m^*, and radius a. Reproduced with permission from R. Marschall, *Adv. Funct. Mater*, 2014, **24**, 2421.[11] Copyright 1991 American Chemical Society.

three parameters, C_1–C_3, and the relationship among them can be summarized in Figure 3.16B.

$$\frac{V}{V_0} = C_1 + \frac{C_2}{\left(\sqrt{V_0 m^* a} + C_3\right)^2} \tag{3.12}$$

In Figure 3.16A, the size dependence of ΔE for ZnO nanoparticles is also shown. Since the dielectric constant of ZnO is small, the large shift with V_h is reduced by the coulomb energy E_{Coul} in eqn (3.10). In the case of CdS with $m^* = 0.19$ and $V_0 = 3.6\,eV$, V calculated by eqn (3.12) becomes about half of that calculated by eqn (3.9).[11]

References

1. Calculation with Gaussian 98W program using STO-3 basis set.
2. R. Asahi *et al.*, *Phys. Rev. B: Condens. Matter Mater. Phys.*, 2000, **61**, 7459.
3. P. A. Cox, *The Electronic Structure and Chemistry of Solids*, Oxford Publ., 1987.
4. D. J. Griffiths, *Introduction to Electrodynamics*, Prentice-Hall, 1999.
5. P. Deak, B. Aradi and T. Frauenheim, *J. Phys. Chem. C*, 2011, **115**, 3443.

6. R. Marschall, *Adv. Funct. Mater.*, 2014, **24**, 2421.
7. D. E. Scaife, *Sol. Energy*, 1980, **25**, 41.
8. Y. Inoue, *Energy Environ. Sci.*, 2009, **2**, 364.
9. A. Kuback, M. Fernandez-Garcia and G. Colon, *Chem. Rev.*, 2012, **112**, 1555.
10. A. Kudo and Y. Miseki, *Chem. Soc. Rev.*, 2009, **38**, 253.
11. Y. Nosaka, *J. Phys. Chem.*, 1991, **95**, 5054.
12. B. Enright and D. Fitzmaurice, *J. Phys. Chem.*, 1996, **100**, 1027.
13. C. Kormann *et al.*, *J. Phys. Chem.*, 1988, **92**, 5196.
14. H. Lin *et al.*, *Appl. Catal., B*, 2006, **68**, 1.

4 Principles of Photoelectrochemistry

4.1 Electron Transfer Reaction at the Solid–Liquid Interface (Electrochemistry)

Electron transfer reactions which take place at the surface of photocatalysts proceed based on electrochemistry in solution. Electrochemistry covers a wide range of various fields such as electric batteries, electrolysis, plating, chemical sensors and bio- and electrophenomena. In general, the electrochemical reactions have the following three superior characteristics.

1. The chemical reaction can be basically controlled by the applied voltage.
2. The reaction rate can be measured as an electric current, because the reaction proceeds with the movement of electrons.
3. The functions of solid catalysts can be imposed on the electrodes because the reactions proceed over the solid surfaces.

The photocatalytic reaction is a kind of electrochemical reaction in terms of that the reactions involve the charge transfer and that the function of the catalyst is realized on the solid surface. In particular, the photocatalytic reactions in solution are considered to follow the principles of electrochemistry, and the photocatalytic properties have been investigated by using electrodes of the photocatalytic semiconductors. The way in which electrochemistry is related to photocatalytic reactions will be briefly described.

Introduction to Photocatalysis: From Basic Science to Applications
By Yoshio Nosaka and Atsuko Nosaka
© Yoshio Nosaka and Atsuko Nosaka, 2016
Published by the Royal Society of Chemistry, www.rsc.org

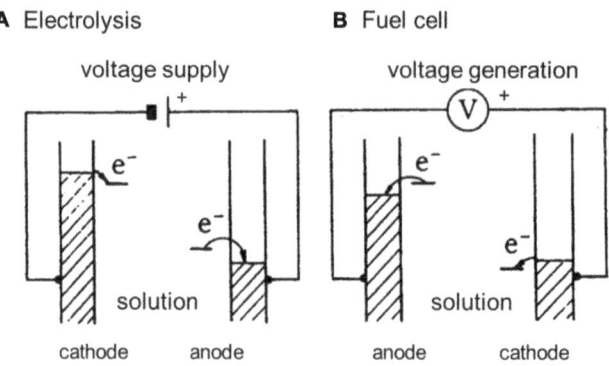

Figure 4.1 The energy levels for electrolysis and the fuel cell.

The reaction familiar with electrochemistry would be the electrolysis of water. In Figure 4.1A and B, the electrolysis of water and the movement of electrons in a fuel cell are compared in terms of energy levels. For the electrolysis of water, as shown in Figure 4.1A, on the application of a positive voltage, the electrode is polarized to positive, and the electronic potential against electrons decreases. Accordingly it enables the electron transfer from the molecules in the solution to the electrode, to cause a positive current with the oxidation of the molecules. The potential of another electrode increases, and the electron can transfer to a molecule in the solution, to cause negative current at this electrode. The reactions in which electrons are released from molecules are called oxidation reactions or anodic processes, while the reactions in which the electrons transfer to a molecule are called reduction reactions or cathodic processes.

The reactions for the electrolysis of water and for fuel cells of hydrogen burning are depicted in Figure 4.2A and B, respectively. Because it would be rather confusing if the electrodes were called positive and negative poles, from now on the electrodes from which electrons are released will be called cathodes, while the electrodes into which electrons enter will be called anodes. Therefore, in fuel cells, the cathode at which O_2 is reduced becomes the positive pole. For the plots of the current against the potential shown in the lower part of Figure 4.2, the right direction of the electric potential and the upper direction of the current are taken as positive. The directions of the flows for electric current (i) and electrons (e^-) are opposite as illustrated in the upper part of Figure 4.2.

The structure of the electrode surface in atomic scale is considered significantly inhomogeneous due to the adsorption of the solvent (water) molecules and ions in the solution as illustrated in Figure 4.3A.[1]

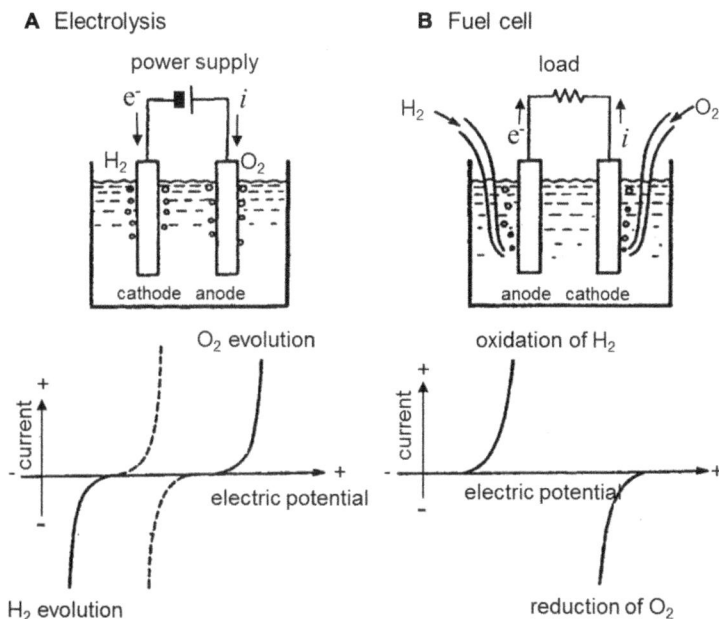

Figure 4.2 Drawings of the model experiments (upper part) for (A) electrolysis and (B) fuel cell, and the electric current-potential curves (lower part). e^-: electron, i: electric current.

The surface of the metal oxide is covered with OH or O^-, then hydrated cations are adsorbed. The layer of cations is called an outer Helmholtz plane or layer. As shown in Figure 4.3B, the layer where the orientation of the solvent molecules is affected by the surface charge is called a Gouy–Chapman layer, and that from the surface to the Gouy–Chapman layer is called an electric double layer. The layer about 10 μm distant from the electrode is not affected by the convection of the solution and is called a diffusion layer.[2] For the reactions on the electrode surface, three processes take place; firstly a reactant R is adsorbed on the electrode surface by diffusion and secondly it is converted to a product P in the adsorbed state by an elementary surface reaction, then it diffuses into the solution. Therefore, besides the rate of the elementary surface reaction, the adsorption and the desorption rates of the reactants in the Helmholtz layer, and the mass transfer rate to the diffusion layer also contribute to the reaction rate at the solid–solution interface.

4.2 Redox Potentials

The potential at which the oxidation and the reduction of molecules or ions take place is generally called a standard redox potential, E^0.

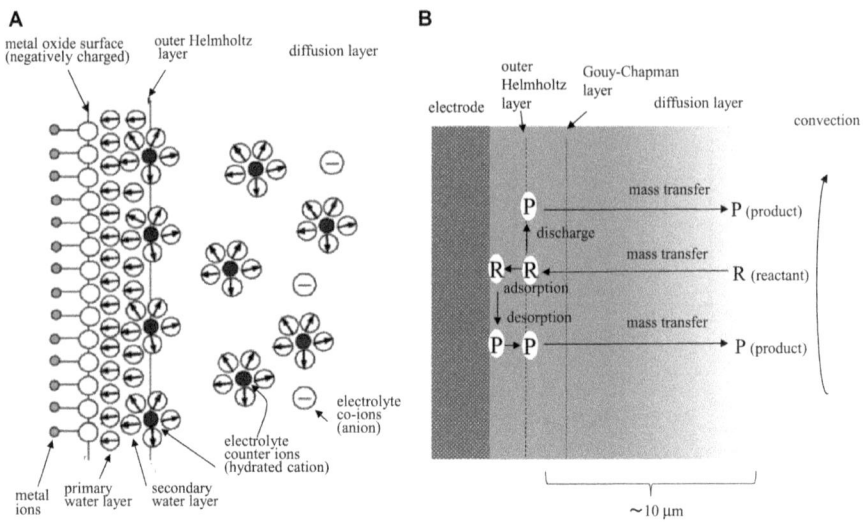

Figure 4.3 (A) Model of the adsorbed molecules in the vicinity of the metal oxide surface. Reprinted from G. D. Panagiotou, T. Petsi, K. Bourikas, C. S. Garoufalis, A. Tsevis, N. Spanos, C. Kordulis and A. Lycourghiotis, Mapping the surface (hydr)oxo-groups of titanium oxide and its interface with an aqueous solution, *Adv. Colloid Interface Sci.*, **142**, 20–42. Copyright (2008) with permission from Elsevier. (B) Reaction processes of the electrochemical reaction in the vicinity of the electrode.

The redox potentials for several compounds are shown in Figure 4.4. The order of the redox potential of metal ions to metal is described as the ionization tendency. For electrochemistry, the potential is taken as an energy axis and the standard hydrogen electrode (SHE, or normal hydrogen electrode, NHE) is taken as the origin of the coordinate axis. Conventionally, for the potential, the upper direction is taken as negative while the lower direction is taken as positive. These are reverse directions to usual graphs because they display potential. A substance that possesses mass on earth at a higher position has a higher potential energy. When it falls from the position, it releases the energy. Similarly, because electrons have negative charge, the position which has more negative potential has higher potential energy. For this reason, the potential is described so that the upper direction becomes negative. Recall that the relationship between energy E and electric potential U is expressed by $E = -eU$, with elementary charge e.

Here, it is noticed that the standard redox potential describes the potential at which the reaction involving electrons as a reactant reaches the equilibrium. Therefore, the potential at which the actual electron current is caused by the electrode reaction is different from

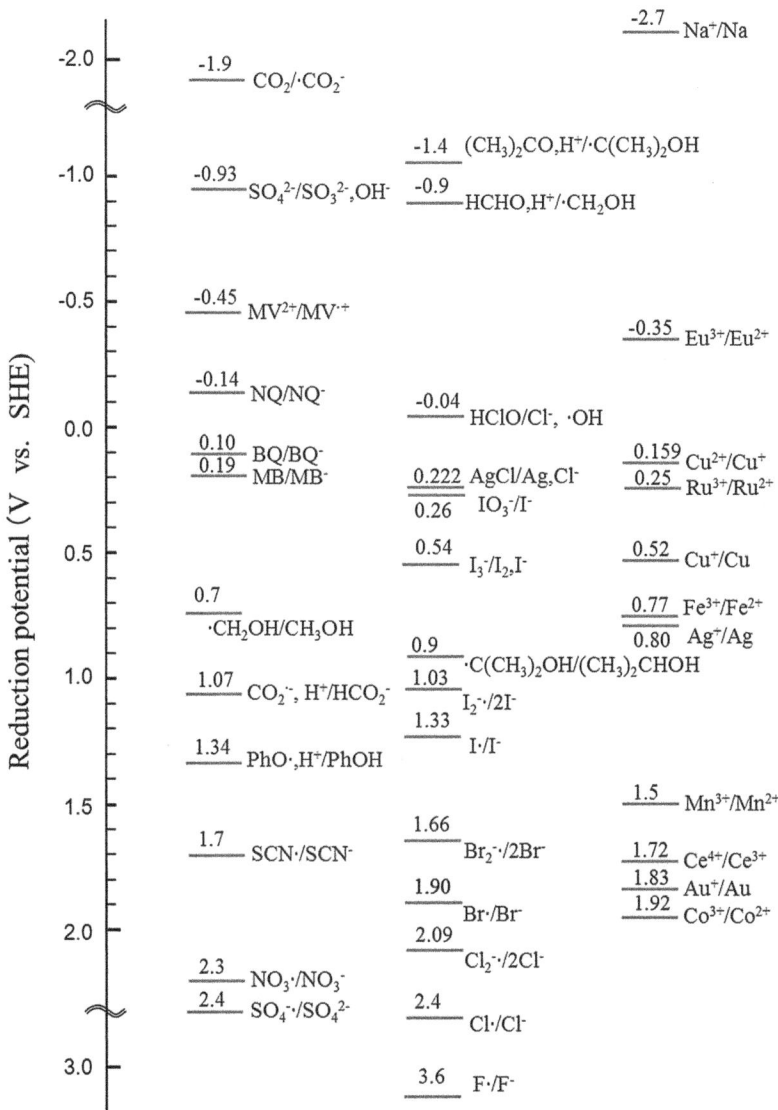

Figure 4.4 One-electron redox potentials for various compounds. MV: methylviologen, BQ: benzoquinone, NQ: naphthoquinone, MB: methylene blue, PhOH: phenol. The data were taken from ref. 3 and 4.

the standard redox potential. It means that a standard redox potential just indicates the change of free energy ΔG^0 of the reaction, where ΔG^0 is related to the difference of E^0 as follows.

$$\Delta G^0 = -nF\Delta E^0 \qquad (4.1)$$

Here, F is the Faraday constant (charge of 1 mole electrons, eN_A), and n is the number of electrons involved in the reaction.

The activation energy necessary for electrode reactions, which can be measured as the potential difference from the redox potential, is called over potential. Because electrodes have a function as catalysts to decrease the activation energy, each electrode and substance has a different over potential. The low over potential means that due to the rapid reaction on the surface, the electrode reactions take place even near the equilibrium potential. In this case, the current at the equilibrium potential, which is called exchange current, becomes a measure of the catalytic activity.

Let's look at the electrode reactions further in detail by taking the electrolysis of water as an example. At the cathodes provided with negative voltage (negatively polarized cathode), the following reaction may take place.

$$2H_2O + 2e^- \rightarrow H_2 + 2OH^- \quad E^0 = -0.828 \text{ V} \tag{4.2}$$

On the other hand, for the positively polarized anode, the following reaction may proceed.

$$4OH^- \rightarrow O_2 + 2H_2O + 4e^- \quad -E^0 = -0.401 \text{ V} \tag{4.3}$$

The addition of these two equations leads to the reaction of the electrolysis of water. By summing eqn (4.2) and (4.3), eqn (4.4) is obtained.

$$2H_2O \rightarrow 2H_2 + O_2 \quad \Delta E^0 = -1.229 \text{ V} \tag{4.4}$$

A standard state for the potential of the semiconductor electrode is usually pH $= 0$, *i.e.*, proton (hydrogen ion) concentration (strictly speaking, activity) is 1 mole, or $[H^+] = 1$ M. For this case, the reduction reaction of water is expressed as

$$2H^+ + 2e^- \rightarrow H_2 \quad E^0 = 0.000 \text{ V} \tag{4.2'}$$

while the oxidation reaction is expressed as

$$2H_2O \rightarrow O_2 + 4H^+ + 4e^- \quad -E^0 = -1.229 \text{ V} \tag{4.3'}$$

The summation of eqn (4.2′) and (4.3′) leads to eqn (4.4).

Thus, the redox potential changes by $-0.0591 \times$ pH along with pH according to the Nernst equation. The factor of 0.0591 is obtained by calculating $\ln(10) \times (k_B T/e)$ at $T = 293$ K. The pH dependences of one-electron redox potential[3] for water and oxygen are shown in Figure 4.5A. In this figure the energy position of conduction band and

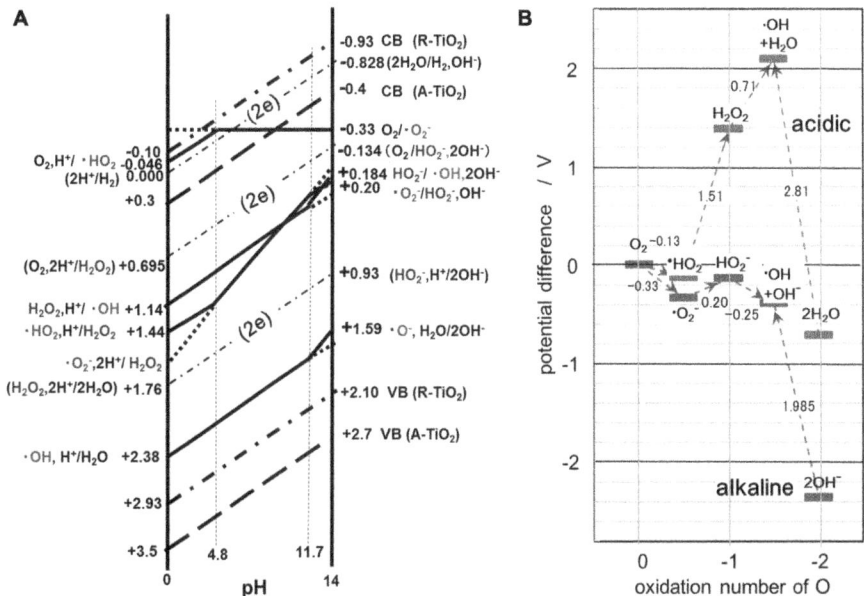

Figure 4.5 (A) pH dependence of one-electron redox of H_2O, H_2O_2, and O_2 with the edges of the conduction band (CB) and the valence band (VB) of TiO_2. Dotted slash line shows two-electron (2e) process. (B) Three-step reduction of O_2 and one-step oxidation of H_2O to OH radical are shown as the potential difference for acidic and alkaline solution.
The data were taken from ref. 3.

valence band of anatase $(A\text{-}TiO_2)^5$ and rutile $(R\text{-}TiO_2)^6$ calculated from literature are also shown because they also depend on pH. For the estimation of band energy, see Section 4.3.2. The band potential of metal oxides becomes higher with an increase of pH since the surface is covered with OH groups which can be ionized. Since the band potentials shift equally to the redox of water, the electric voltage necessary for the photocatalytic water decomposition with metal oxide semiconductors does not depend on pH.

By using eqn (4.1), the change of the free energy ΔG^0 for the electrolysis reaction of water eqn (4.4) with four electrons is estimated to be $-4 \times 96\,500$ C $mol^{-1} \times (-1.229$ V$) = 474$ kJ mol^{-1}. Although the difference of redox potential, ΔE^0, is 1.229 V, the reaction does not proceed with this voltage but the voltage four times higher than 1.229 V, which corresponds to ΔG^0, must not be necessarily applied for the four-electron reaction. Thus, ΔE^0 was just calculated as the change of the free energy per electron required for this four-electron-reaction.

For the actual water electrolysis, the selection of the reaction process employing four electrons for the individual elementary reaction and the decrease of the over potential become important issues. Therefore, one-electron redox potentials are mainly shown in Figures 4.4 and 4.5. In Figure 4.5B the step of the reduction of O_2 to H_2O is shown by the standard potential for acid and alkaline solutions.[3] O_2 is reduced to $^\bullet O_2^-$, H_2O_2, and OH radical, and then becomes H_2O along with the change of oxidation number of oxygen from 0 to -2. For the water oxidation, the reverse process proceeds. This figure shows that the oxidation and reduction energy depends largely on the pH of the solution.

4.3   Semiconductor Photoelectrochemistry

4.3.1   Honda–Fujishima Effect

As for the research adopting semiconductors as electrodes, the research by Gerischer's group preceded. They used the semiconductor for electronic materials such as ZnO as an electrode.[7] However, in 1969 Fujishima and Honda reported that by using the electrode adopting a single crystalline TiO_2, on light irradiation, hydrogen gas from the cathode of platinum Pt, and oxygen gas from the anode of TiO_2 were generated as shown in Figure 4.6.[8] As described in Section 1.6, the report was a landmark in terms of the first finding of the metal oxide semiconductor from which the metal did not dissolve in the solution with anodic current.

Figure 4.7 shows the energy level of a photoelectrochemical cell adopting TiO_2 and platinum electrodes. The Fermi level E_F of Pt without potential application is known as the work function, which is the energy difference between the vacuum state and the occupied electrons. The energy level of SHE is known to locate at 4.5 eV below the vacuum. Because the Fermi level of n-type TiO_2 semiconductor is higher than that of Pt metal, on connecting two electrodes without applying potential, electrons transfer from TiO_2 to Pt so as to even the two Fermi levels. In general, the electron density of semiconductors is much lower than that of metals. Therefore, the Fermi level of the metal scarcely changes on the electron transfer, while the Fermi level of the semiconductor moves to the position of the metals. In this situation, the potential at the surface of the semiconductor is determined by the materials and surface charge, because there are enough ions to supply charges for the semiconductor in electrolyte solution. As a result, the bending of the band (potential gradient) is generated as shown in Figure 4.7A with solid

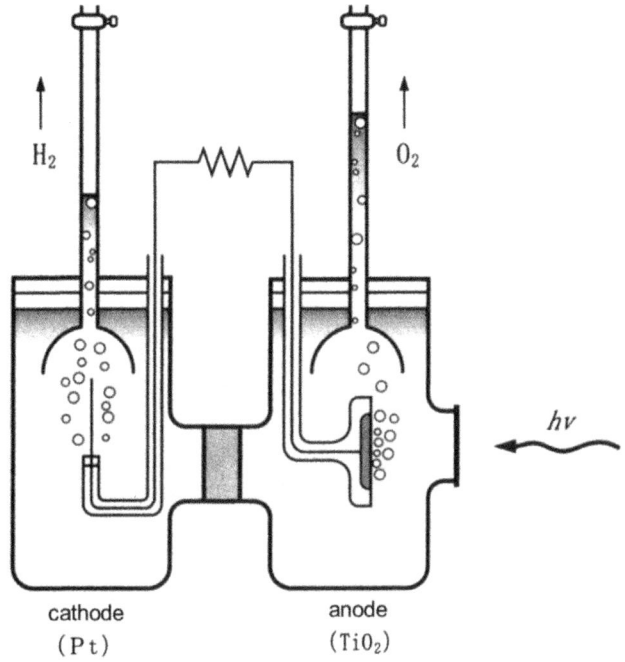

Figure 4.6   A photoelectrochemical battery adopting TiO$_2$ as an anode.

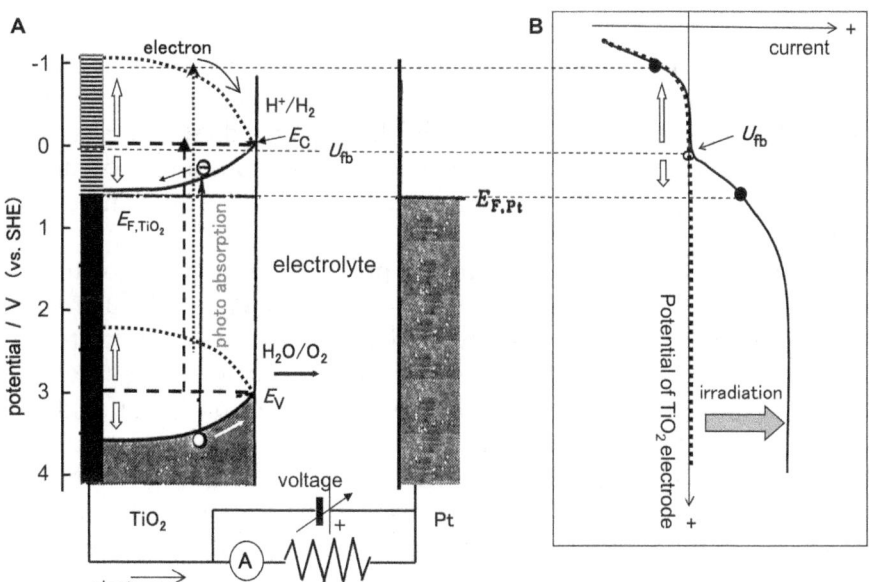

Figure 4.7   Explanation of Honda–Fujishima effect by presenting the relation between (A) the energy level for the titanium oxide semiconductor and (B) the observation of photocurrent.

curves. This is called a space charge layer or a depletion layer as will be discussed later in detail.

Figure 4.7B shows the potential current curve which is oriented to fit the direction of potential in Figure 4.7A. When the valence band holes and conduction band electrons are generated by irradiating such a TiO_2 electrode, electrons fall down along the gradient of the conduction band to flow *via* a lead wire into the Pt counter electrode. On the other hand, holes produced in the valence band move to the upper along the gradient to reach the surface where the water could be oxidized. When such electron transfer reactions take place on the surface by absorbing light, electric current is observed. This current is called photocurrent, which is the evidence of redox reactions on the TiO_2 surface. On the potential application to the TiO_2 electrode, the photocurrent changes like a solid line in Figure 4.7B. The current marked with a closed circle ● in Figure 4.7B corresponds to that observed with the potential described in Figure 4.7A. The current does not increase linearly on the application of high positive potential but becomes constant at a certain potential. This means that the diffusion of the substance to and from the semiconductor surface is the rate-control process.

When negative potential relative to the Pt electrode is applied, the potential gradient diminishes as shown by the broken line on the left side of Figure 4.7A since the Fermi level of titanium oxide (E_F, TiO_2) rises. Namely, because the band becomes flat, the potential is called a flat band potential U_{fb}, at which the current does not flow. With further application of the potential of TiO_2 in the negative direction, electrons flow in the reverse direction, even without photoabsorption, and water is reduced at the surface of the electrodes, that is, the reduction current starts to flow.

The behavior of the current without photoirradiation is shown with a dotted line in the potential-current curve in Figure 4.7B. At the negative potential of TiO_2, the reductive current flows similar to the case of light irradiation. However, for the application of positive potential to lower E_F, TiO_2, the molecule in the solution is not oxidized unless the light is irradiated. Therefore, electrons do not transfer to the TiO_2 electrode. It means that the oxidation current does not flow. Such principle that on the application of the potential the current flows only to one direction is similar to that of a diode (a rectifier) with a Schottky barrier.

Let's consider the case of water containing organic molecules, such as alcohols. In general, the chemical bond forming molecules consists of a pair of electrons. Therefore, the molecule oxidized by the abstraction of one electron on the surface of TiO_2 is so unstable to release easily another electron. Consequently, one electron is sometimes injected to the electrode from the oxidized molecule. In the presence of such molecules, the oxidative photocurrent becomes twice as large. This effect is called a current doubling effect. For photocatalytic reactions, the enhancement of reactions due to the current doubling effect may be possible.

4.3.2 Flat Band Potential and Mott–Schottky Plot

As stated above, the potential in TiO_2 near the surface in Figure 4.7A is similar to the Schottky barrier in semiconductor physics. To understand the potential gradient further, the analysis method by Mott–Schottky plot will be described from now on. In the case of a semiconductor electrode, because the electrolyte concentration in solution is high, the potential at the surface is retained when electrons move from semiconductor to solution. As shown in Figure 4.8, on the movement of electrons, positive charges remain in the semiconductor, leading to lower potential. The layer where the effect of

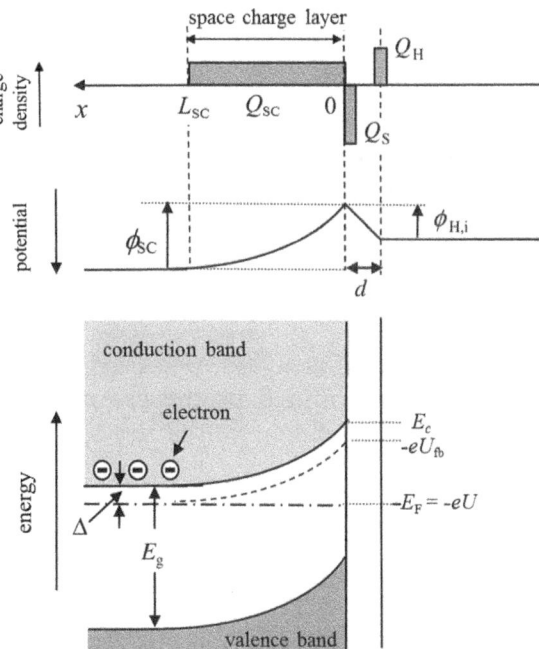

Figure 4.8 Electric potential and energy of space charge layer near the surface of n-type semiconductor with adsorption of anions. Reproduced from ref. 9 with permission from the Electrochemical Society of Japan.

positive charge is caused is called a space charge layer. Here, negative charges Q_S may be caused by the adsorption of negative ions, and the positive charge in the solution may form inner Helmholtz layer Q_H to compensate the charge.[9,10]

Let's calculate the thickness of the space charge layer, L_{SC}. The shape of the potential at distance x from the surface, $\phi(x)$, obeys the Poisson–Boltzmann equation eqn (4.5),

$$\frac{\mathrm{d}^2\phi(x)}{\mathrm{d}x^2} = -\frac{\rho}{\varepsilon_0\varepsilon} \tag{4.5}$$

where ε_0 is the permittivity of vacuum, ε is the dielectric constant (or relative permittivity) of the semiconductor, ρ is charge density and it can be approximated as $\rho = eN_0$, with elementary charge e and the carrier density N_0 in the semiconductor. Eqn (4.5) can be solved under the boundary condition of $\mathrm{d}\phi(x)/\mathrm{d}x = 0$ and $\phi(x) = 0$ at $x = L_{sc}$.

$$\phi(x) = -\frac{eN_0}{2\varepsilon_0\varepsilon}(x - L_{SC})^2 \tag{4.6}$$

Since the potential difference can be given by $\phi_{sc} = -\phi$ (0), the thickness is expressed by eqn (4.7).

$$L_{SC} = \sqrt{\frac{2\varepsilon_0\varepsilon\phi_{SC}}{eN_0}} \tag{4.7}$$

Though it is not easy to measure N_0 of TiO_2 particles, the actual measured value of the thin layer of the particle used for the dye-sensitized electrode is $(0.2 \sim 1.0) \times 10^{17}\,\mathrm{cm}^{-3}$.[11] For anatase TiO_2, with the known values of $\varepsilon = 12$, $\phi_{sc} = 1.0$ V, and $N_0 = 10^{17}\,\mathrm{cm}^{-3}$, the thickness of the space charge layer can be calculated to be $L_{sc} = 115$ nm from eqn (4.7).

Next, let's refer to the flat band potential U_{fb}. As shown in Figure 4.8, since E_F is determined by the external potential U, the difference of the potentials which causes the space charge layer is expressed by eqn (4.8).

$$\phi_{SC} = U - U_{fb} - \frac{k_BT}{e} \tag{4.8}$$

The electric charge, Q_{sc}, of the space charge layer is expressed by eqn (4.9).

$$Q_{SC} = eN_0L_{SC} \tag{4.9}$$

From eqn (4.7)–(4.9), eqn (4.10) is derived.

$$Q_{SC} = \sqrt{2\varepsilon_0\varepsilon eN_0(U - U_{fb} - k_BT/e)} \tag{4.10}$$

Q_{sc} can be obtained by measuring differential capacitance C_{SC}, which is obtained with eqn (4.11).

$$C_{SC} = \frac{dQ_{SC}}{dU} \tag{4.11}$$

By substituting eqn (4.10) for eqn (4.11), eqn (4.12) is derived.

$$\frac{1}{C_{SC}^2} = \frac{2(U - U_{fb} - k_BT/e)}{\varepsilon_0\varepsilon e N_0} \tag{4.12}$$

Because the capacitance C_{sc} can be measured with an impedance meter, U_{fb} and N_0 can be obtained experimentally by measuring the left part of eqn (4.12) as a function of U. The plot of eqn (4.12) is called a Mott–Schottky plot and is utilized to determine the band position of semiconductors after fabricating an electrode.

To obtain the energy of the conduction band bottom, E_c, from U_{fb}, the difference Δ between U_{fb} and E_c shown in Figure 4.8 must be estimated. Δ can be estimated with eqn (4.13),

$$\Delta = k_BT\ln\left(\frac{N_C}{N_0}\right) \tag{4.13}$$

where N_c is the effective density of conduction band state (DOS between E_c and $E_c + k_BT$), which is about $10^{19}\,cm^{-3}$. For rutile TiO_2, the reported value is $N_c = 2.5\times10^{19}\,cm^{-3}$.[12]

The measurements of U_{fb} for rutile and anatase TiO_2 have been reported previously.[13] In this report, the U_{fb} of anatase and rutile TiO_2 were reported to be -0.4 and -0.20 V (*vs.* SCE) in 1 M H_2SO_4, respectively,[13] where the potential of SCE (saturated calomel electrode) is 0.244 V relative to that of SHE at 25 °C. Though U_{fb} of anatase is higher (or more negative in potential) than rutile TiO_2 based on the Mott–Schottky plot, it was recently suggested that E_c of rutile should be higher by 0.4 eV than that of anatase, according to state-of-the-art materials simulation and X-ray photoemission experiments.[5] Therefore, the Mott–Schottky plot reported previously for anatase TiO_2 was most likely measured for a higher level of the conduction band which corresponds to the direct bandgap of $E_g = 3.8$ eV. Further discussion on the energy difference between anatase and rutile will be presented in Chapter 7.

As for U_{fb} of rutile TiO_2, recent experimental data[6] show $U_{fb} + k_BT/e = -0.305$ V (*vs.* Ag/AgCl) as an average of (110) and (100) surfaces at pH 1.1 in 0.1 M $HClO_4$ with $N_0 = 8.2\times10^{18}\,cm^{-3}$. This corresponds to $U_{fb} = -0.07$ V (*vs.* SHE) at pH $= 0$ because an Ag/AgCl reference electrode is by 0.199 V more positive than SHE. From these data, $\Delta = 0.029$ V and then E_c of rutile becomes $E_c = -0.10$ V (*vs.* SHE) at pH $= 0$.

For the measurements of differential capacitance, the instrument shown in Figure 4.9 is used,[9] where an impedance analyzer is attached to the electrochemical apparatus adopting the potentiostat with a popular three-electrode cell. The potentiostat regulates the potential of the counter electrode so as to apply the desired potential between the semiconductor electrode and the reference electrode. Minute oscillating voltage from the impedance analyzer is applied on the potentiostat. By detecting the corresponding response as the change of the intensity and the phase of the response current (ac i) and potential (ac U), the differential capacitance is calculated.

As shown in Figure 4.8, when anion Q_s is adsorbed on the surface, the cation gathers on the outer Helmholtz surface to generate the potential difference of ϕ_H. As a result, the potential difference is generated between the surface of the electrode and the bulk solution, suggesting that the measured U_{fb} should shift by ϕ_H. This must be kept in mind on determining U_{fb}.

The above equations are derived for n-type semiconductors and electron carriers. Similar equations can be derived for p-type semiconductors by adopting the carrier density of holes. In this case, the

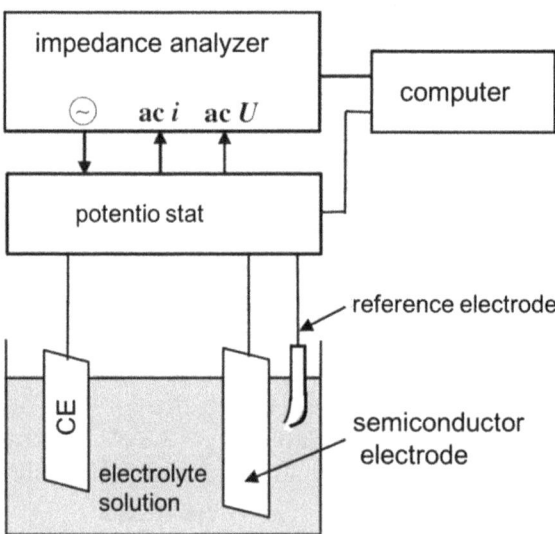

Figure 4.9 Schematic illustration of experimental set-up to measure impedance of semiconductor electrodes. CE is a counter electrode, usually a Pt electrode.
Reproduced from ref. 9 with permission from the Electrochemical Society of Japan.

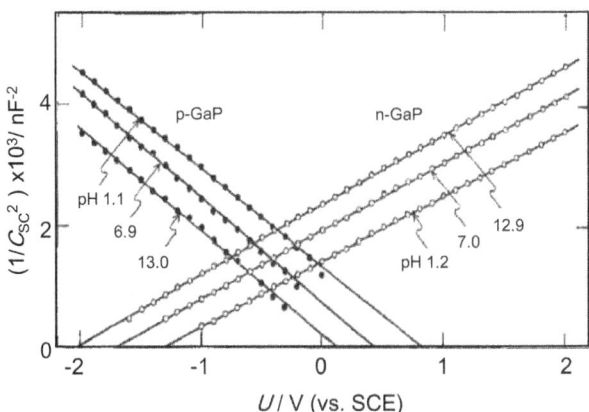

Figure 4.10   Mott–Schottky plots of n- and p-GaP electrodes in 0.1 M
NaOH with the frequency of 1 kHz.
Reproduced from ref. 9 with permission from the Electro-
chemical Society of Japan.

sign of eqn (4.12) becomes negative. In Figure 4.10, an example of a
Mott–Schottky plot for n- and p-types of GaP semiconductor measured in solutions under various pH is shown.[9] Not only for metal
oxides but also for phosphorus compounds, when they possess the
chemical structures which can cause the proton dissociation, with
increasing pH the negative charge on the surface increases resulting
in the change of U_{fb} by $-0.059 \times pH$ according to the Nernst equation.

4.3.3   Energy Structure of Semiconductor Particles

The internal energy structure of a semiconductor particle is shown in
Figure 4.11, where the Fermi level E_F of the semiconductor attains in
the equilibrium with the energy level of the solution. The energy level
at the surface is determined by the substance and independent of the
energy band levels inside the semiconductor. Therefore, when electrons move to the solution to be equal to the potential of the solution,
the position of the conduction band bottom goes down to near the E_F
level forming a space charge layer.

When a particulate semiconductor is treated as a sphere of radius a,
a Poisson–Boltzmann equation of the spherical coordinate, eqn
(4.14), must be solved to calculate the potential ϕ of the space charge
layer at the position r from the particle center.

$$\frac{1}{r^2}\frac{\partial}{\partial r}\left(r^2\frac{\partial\phi}{\partial r}\right) = \frac{eN_0}{\varepsilon\varepsilon_0} \qquad (4.14)$$

Under the condition of $\dfrac{\partial \phi}{\partial r} = 0$ at $r = a - L_{SC}$, the solution is given by eqn (4.15).[14]

$$\phi(r) = \left(\frac{eN_0}{6\varepsilon_0\varepsilon}\right)(r - a + L_{SC})^2\left(1 + \frac{2(a - L_{SC})}{r}\right) \tag{4.15}$$

When the radius of the particle is larger than the thickness of the space charge layer $(a \gg L_{sc})$, the decrease of the potential ϕ_0 becomes equal to ϕ_{sc} as shown in Figure 4.11A.

For nanoparticles with small particle size $(a < \sqrt{3}L_{sc})$, as shown in Figure 3.11B the whole particle is contained in the space charge layer and the decrease of the potential at the particle center ϕ_0 can be calculated with eqn (4.16).

$$\phi_0 = \frac{a^2 eN_0}{6\varepsilon_0\varepsilon} \tag{4.16}$$

When the particle diameter of TiO_2 semiconductor is $2a = 20\,\text{nm}$, the decrease of the potential at the particle center ϕ_0 can be calculated with eqn (4.16) to be about 0.3 mV under the assumption of $N_0 = 10^{17}\,\text{cm}^{-3}$. Therefore, the potential drop at the center of the particle may be ignored for nanoparticles.

When the surface of the semiconductor nanoparticle is modified by depositing a metal cluster, at the interface the potential barrier may

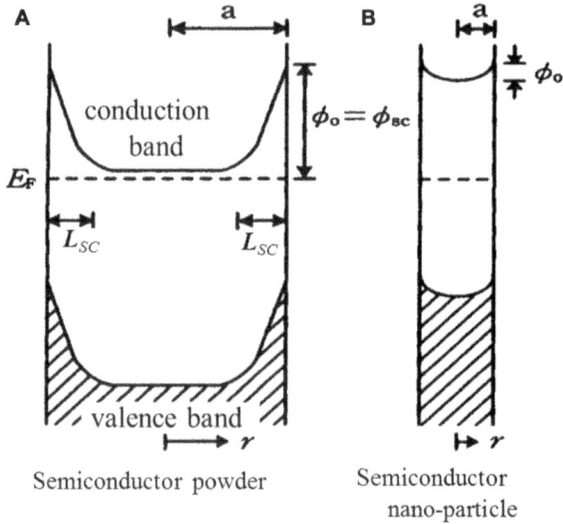

Figure 4.11   Cross section of the energy band potential in semiconductors of different sizes of particles.

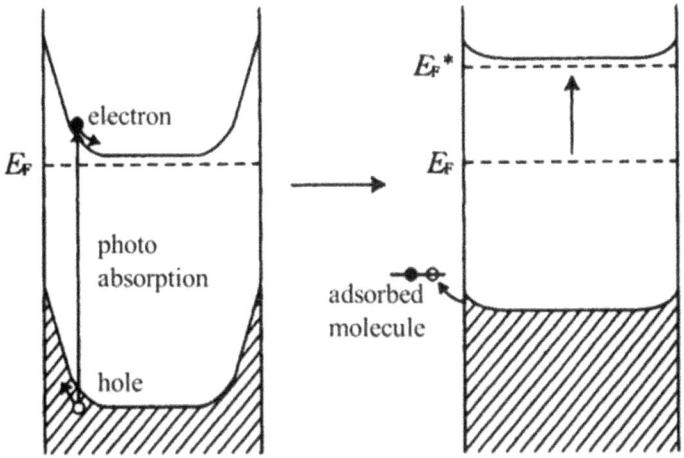

Figure 4.12 Moving-up of the Fermi level from E_F to E_F^* of a semiconductor by the accumulation of electrons with photoabsorption.

be generated in some cases or in other cases ohmic contact may be generated without a barrier. Then, a large potential difference is considered to be formed between the conduction band of the semiconductor and the Fermi level of the metal cluster (see Section 7.6).[15]

As shown in Figure 4.12, when a semiconductor particle with a space charge layer absorbs light, the generated holes move to the surface along the electric gradient and oxidize absorbents. On the other hand, the electrons gather towards the center along the electric gradient and are accumulated in the particles. Therefore, the Fermi level elevates and the space charge layer decreases.

4.4 Dye-sensitized Solar Cell

In the case of popular practical solar cells, the electric field gradient is generated at the junction of n- and p-type semiconductors, and the photoabsorption causes the charge separation at the junction to provide electric currents. The principle of the Honda–Fujishima cell shown in Figure 4.7 is similar to that for the case of a solar cell. However, for water decomposition reactions, because of the high over potential to precede the reaction, the additional potential by a battery is required besides the photoirradiation.

When using TiO_2 for a solar cell, the low absorption of the solar light is a serious problem. To overcome this problem, the research on

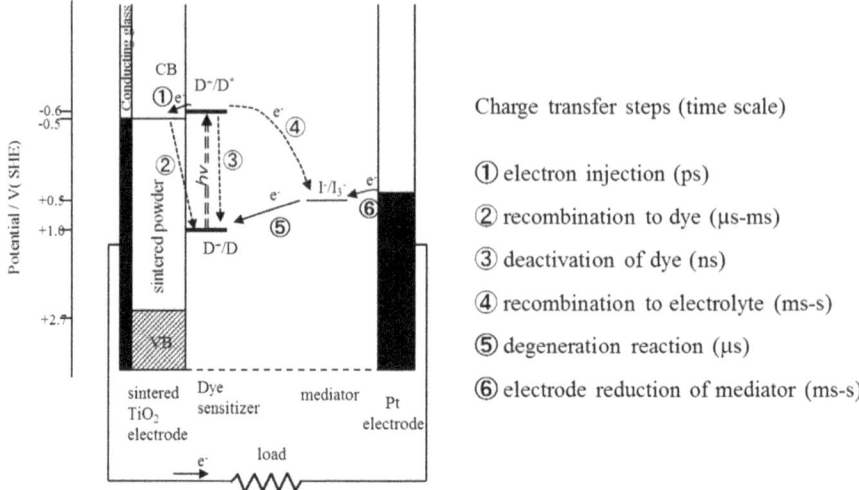

Figure 4.13   Schematic illustration for the principle of a dye-sensitized solar cell with undesired charge transfer path (dotted line) and the time scales of each step.

the photosensitized solar cells in which an electron was injected from an excited molecular dye was reported in late 1970. Grätzel *et al.* succeeded in the effective photo-electron conversion by adopting a porous TiO_2 film electrode deposited by a dye.[16] The electron transfers from the photo-excited dye D* to the TiO_2 electrode and then the dye itself becomes the oxidant D^+ as depicted in Figure 4.13. The oxidized dye D^+ reverts to the original D by oxidizing an iodide ion I^- in the solution. I^- becomes the iodide molecule I_2 or I_3^- by the oxidization, and is reduced at the counter electrode to revert to the original I^-. The undesired electron transfer process and the time scale of electron transport are discussed in the report[17] and summarized in Figure 4.13. If the conversion efficiency of solar light were comparable to that of the solar cells with silicon semiconductors, it would be expected that a part of the present silicon solar cells could be replaced by the dye-sensitized solar cells because fabrication with TiO_2 is less expensive.

For a practical use of a dye-sensitization solar cell, the instability of I^- solution and organic dye molecules are the problems. To avoid the degradation of electrolyte solution, a solid state electrolyte has been suggested.[18] The dye molecules could be replaced by narrow bandgap semiconductors.

References

1. G. D. Panagiotou, K. Bourikas, *et al.*, *Adv. Colloid Interface Sci.*, 2008, **142**, 20.
2. H. Kita and K. Uosaki, *Denkikagaku no Kiso*, Gihodo Shuppan, Co. Ltd., Tokyo, 1990.
3. *Standard Potentials in Aqueous Solution*, ed. A. J. Bard, R. Parsons and J. Jordan, Marcel Dekker, New York, 1985.
4. P. Wardman, *J. Phys. Chem. Ref. Data*, 1989, **18**, 1637.
5. D. O. Scanlon, J. Buckeridge, *et al.*, *Nat. Mater.*, 2013, **12**, 798.
6. E. Tsuji, K. Fukui, A. Imanishi, *et al.*, *J. Phys. Chem. C*, 2014, **118**, 5406.
7. H. Gerischer, *J. Electrochem. Soc.*, 1966, **113**, 1174.
8. A. Fujishima and K. Honda, *Nature*, 1972, **238**, 37.
9. Y. Nakato, *Electrochemistry*, 2014, **82**, 507.
10. J. O'M. Bockris and S. U. M. Khan, *Surface Electrochemistry*, Plenum Press, New York, 1993.
11. N. Kopidakis, E. A. Schiff, *et al.*, *J. Phys. Chem. B*, 2000, **104**, 3930.
12. H. Tang *et al.*, *J. Appl. Phys.*, 1994, **75**, 2042.
13. L. Kavan *et al.*, *J. Am. Chem. Soc.*, 1996, **118**, 6716.
14. W. J. Albery and P. N. Bartlett, *J. Electrochem. Soc.*, 1984, **131**, 315.
15. Y. Nosaka, K. Norimatsu and H. Miyama, *Chem. Phys. Lett.*, 1984, **106**, 128.
16. B. O'Regan and M. Grätzel, *Nature*, 1991, **353**, 737.
17. M. Grätzel and J. S. Durrant, in *Series of Photoconversion of Solar Energy*, ed. M. D. Archer, Imperial College Press, 2008, vol. 3, ch. 8, p. 503.
18. J. Wu *et al.*, *Chem. Rev.*, 2015, **115**, 2136.

5 Photocatalyst Surface and Active Species

5.1 Structure of TiO$_2$ Surface and the Adsorbed Water

The photocatalytic reactions proceed on the solid surface as is well known. Therefore, it is important to know the characteristics of the solid surfaces or to understand the photocatalytic reaction mechanisms. Let's take a glance at the surface characteristics of TiO$_2$, which is practically used as a photocatalyst. One should keep in mind that the surface structures change with the variations of the surrounding atmosphere and/or the environments. The surface atmospheres of TiO$_2$ may be categorized into four atmospheres, *i.e.*, vacuum, air, aqueous solution, and organic solvents, as depicted schematically in Figure 5.1. Because Ti^{4+} of TiO$_2$ takes 6-coordination while O^{2-} takes 3-coordination in the crystal, they can be depicted in a diamond shape in an expansion plane as shown in this figure.

The photocatalytic reactions have been investigated by exploiting various kinds of experimental methods depending on the individual surface conditions. In the research field of surface science, the photocatalytic reactions can be categorized roughly into the reactions under ultra-high vacuum. The reaction in the aqueous solution is in the research field of electrochemistry, and that in the organic solvents is in the research fields of photochemistry and synthetic chemistry. Because photocatalysts are practically used mostly under atmospheric conditions, research on photocatalysis tends to be concentrated under conditions of humidified air.

Introduction to Photocatalysis: From Basic Science to Applications
By Yoshio Nosaka and Atsuko Nosaka
© Yoshio Nosaka and Atsuko Nosaka, 2016
Published by the Royal Society of Chemistry, www.rsc.org

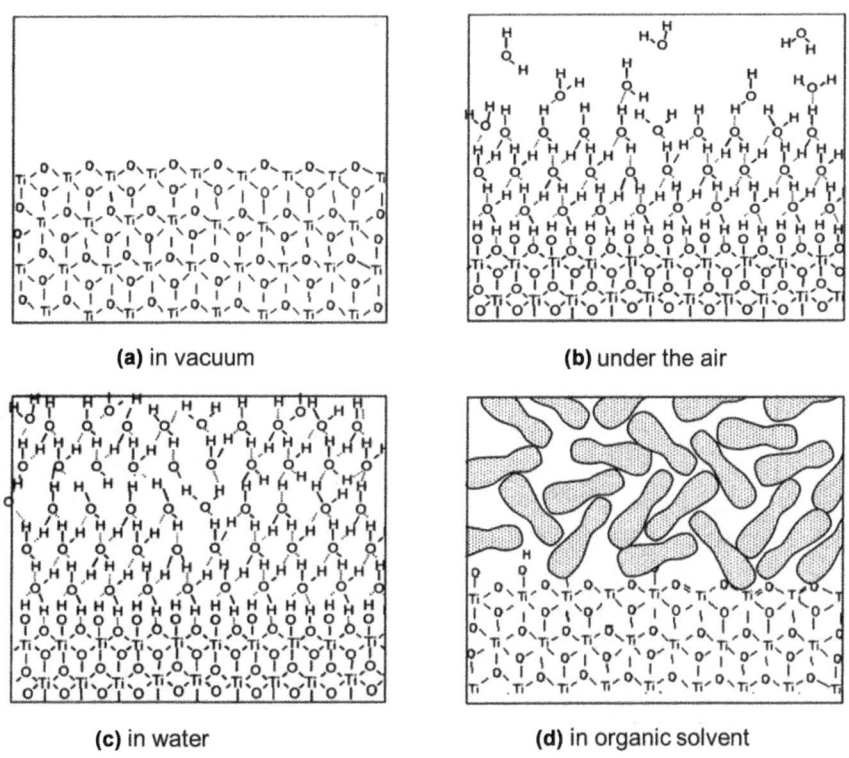

(a) in vacuum   **(b)** under the air

(c) in water   **(d)** in organic solvent

Figure 5.1 Schematic illustration for the TiO_2 surface (a) in vacuum, (b) under moist air, (c) in water, and (d) in organic solvent.

5.1.1 The Surface under Gases

The reaction mechanisms on the solid surface have been investigated by adopting the conventional analytical techniques of the solid surface such as XPS (X-ray Photoelectron Spectroscopy), TPD (Temperature Programmed Desorption), STM (Scanning Tunneling Microscopy) and HREELS (High Resolution Electron Energy Loss Spectroscopy). However, on the application of these techniques, the existence of the molecules besides the desired adsorbents which are involved in the photocatalytic reactions on the solid surface might disturb the measurements. For this reason, in most cases, the observations of the structure and electric states on the solid surface are carried out under ultra-high vacuum conditions. However, because these experiments are performed on an ideally clean surface, one should be careful that the obtained results actually reflect those for the practical photocatalysts under the atmospheric condition. On the surface science of the metal oxides containing TiO_2, Diebold[1] and

Henderson[2] provide comprehensive reviews in which the research on the adsorbent molecules on the surface of the crystal of rutile TiO_2 are surveyed in detail. The review which is specified in rutile (110) TiO_2 surface with SPM (Scanning Probe Microscopy) is also a good reference.[3]

On the surface of TiO_2, water molecules are usually adsorbed. As will be stated later, water is required for the photocatalytic reactions for environmental clean-up and the absorption of the water plays an important role in the reaction mechanism of the photocatalysis. The adsorbed states of water on the TiO_2 surface under vacuum can be suspected on the basis of the experimental results obtained with TPD. The desorption behaviors of water at the facets of (110) and (100) of rutile TiO_2 crystal are shown in Figure 5.2A.[4] Under such vacuum conditions, water is adsorbed in the form of a molecule. The adsorption at the (100) facet, where the surface energy is low, is so weak that the water molecules are desorbed at the lower temperature in the TPD process. The peak observed at higher temperature is interpreted as waters adsorbed on the surface defects. On a clean crystal surface without defects water is not adsorbed dissociatively, but is dissociatively adsorbed at the region where terrace or point defects exist as shown in Figure 5.2B.[4]

On simultaneous adsorption of water and oxygen molecules on the (110) facet, the oxygen is adsorbed on the surface as an ad-atom and the oxygen defects are restored. Then, the ad-atom causes successively the dissociative adsorption of water. The adsorbed OH group formed

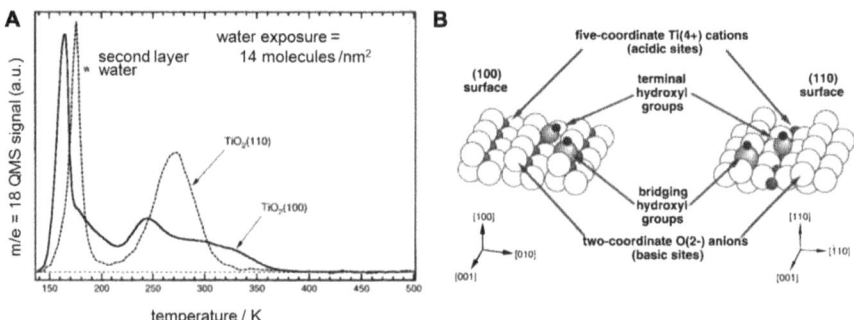

Figure 5.2 (A) Comparison of the temperature programmed desorption (TPD) spectra obtained from water adsorption on different surfaces of rutile TiO_2 single crystals. (B) Schematic representations of terminal and bridged hydroxyl groups on the TiO_2 (110) and TiO_2 (100) surfaces.
Reprinted with permission from M. A. Henderson, *Langmuir*, 1996, **12**, 5093.[4] Copyright 1996 American Chemical Society.

in this way exists on the surface even at 600 K (327 °C) without dissociation. Because water molecules are adsorbed dissociatively on the surface under atmospheric conditions as deduced from the experimental results under the vacuum conditions, it is suspected that quite a number of surface hydroxyl groups exist on the surface of rutile TiO_2.

The correlation between the density of the surface hydroxyl groups of TiO_2 and evacuation temperatures is shown in Figure 5.3. The saturation density of the surface hydroxyl groups is 9–11 nm^{-2} for rutile and 12–14 nm^{-2} for anatase TiO_2, respectively. At 200 °C, most of the hydroxyl groups are eliminated and only one-quarter remain. On further increase in the temperature above 500 °C, the surface hydroxyl groups are completely diminished. After the increase of temperature up to 800 °C, the hydroxyl groups do not recover on reverting the temperature to room temperature and immersing the crystal in water.[5]

For the rutile powder prepared by calcinations at 550 °C, the amount of the adsorbed water remains by 25% at 25 °C after the evacuation but on increase of temperature up to 150 °C water is

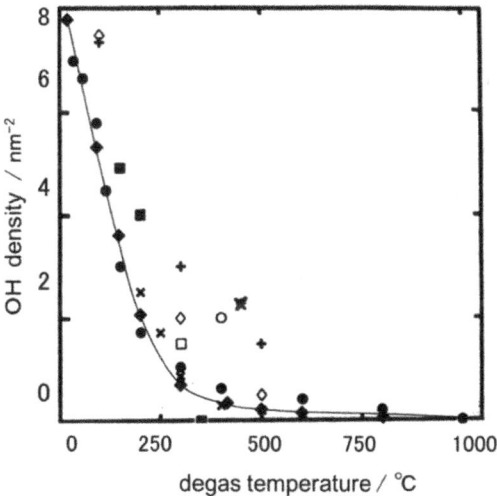

Figure 5.3 Desorption of adsorbed water from TiO_2 powders: Rutile; + active hydrogen analysis, ◇◆ ignition loss, ●x water vapor adsorption, O temperature programmed desorption. Anatase; □ deuterium exchange, and acetylation.[5,6]
Reproduced with permission from Y. Suda and T. Morimoto, *Langmuir*, 1987, **3**, 786. Copyright 1987 American Chemical Society.[6]

completely eliminated. That is, at 150 °C with evacuation, water adsorbed in the form of molecules is desorbed.[6] On further increase of temperature above 350 °C, the released water is formed by combining two surface hydroxyl groups. Thus, the structure of the hydroxyl groups on the surface of TiO$_2$ changes. The solid line in Figure 5.3 indicates the recent data of the continuous change of the amount of the hydroxyl groups against the degas temperature on the rutile TiO$_2$.[6]

Figure 5.3 also shows that the surface hydroxyl groups for anatase TiO$_2$ diminish at lower temperature than those for rutile TiO$_2$. On measurement by IR spectroscopy, it was indicated that the surface hydroxyl groups on anatase TiO$_2$ desorbed irreversibly on drying even at room temperature. On heating at 300–400 °C under vacuum, two kinds of hydroxyl groups, bridged form (Ti$_2$>OH) and terminal form (Ti–OH), are still recognized. However, above 700 °C under vacuum OH groups scarcely remain. On the surface of the anatase after being strongly dehydrated, most of the remaining waters adsorbed on the surface molecularly, but a part of the water molecules adsorbed dissociatively. On repeating the dehydroxylation, the adsorption part on which water can adsorb dissociatively becomes less. This may be attributable to the reconstruction of the atomic configuration on the surface in the process of the dehydroxylation.[7] Thus, taking account of the various exertions which heat treatment causes, one can understand that the last calcination temperatures affect the surface state of the materials.

5.1.2 The Surface under Liquid

In organic solvents, the amount of water that can be adsorbed on the surface is small. Therefore, it is suspected that the surface of TiO$_2$ can be depicted as in Figure 5.1(d). In a solvent like alcohol, the alcohol molecules seem to chemisorb on the Ti atom as a form of alkoxide, just like the adsorbed structure reported under vacuum.

Figure 5.4 shows the zeta potential of TiO$_2$ suspended in water, mixed water–ethanol (1:1), and ethanol as a function of concentration of added NaCl, which was measured to investigate the effect of the addition of NaCl on the surface charge.[8] The zeta potential is measured as a potential at the slipping plane position. The slipping plane is located outside of the outer Helmholtz layer near the solid surface shown in Figure 4.3A. It locates at 1.5 nm and 1 nm distant from the surface in water and in water–ethanol mixed solvent, respectively.[8] Therefore, strictly speaking, the zeta potential is not

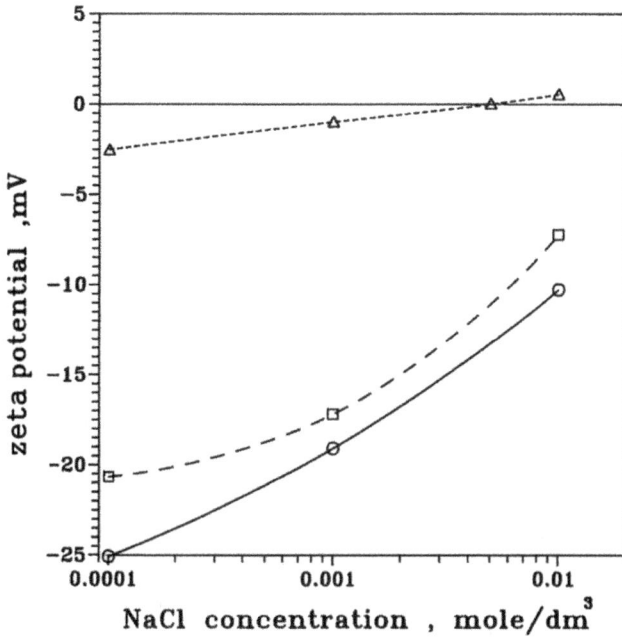

Figure 5.4   The zeta potential of anatase TiO₂ in aqueous (O), mixed aqueous–ethanol (1:1) (□), and ethanol solutions (△) as a function of concentration of NaCl.
Reprinted from W. Janusz, A. Sworska and J. Szczypa, The structure of the electrical double layer at the titanium dioxide/ethanol solutions interface, *Colloids Surf., A*, **152**, 223–233. Copyright (1999) with permission from Elsevier.

the surface potential. But it is considered to be approximately equivalent to the surface potential in the absence of the specific adsorption of ions.

   The isoelectric point is defined as the pH at which the net charge of the surface becomes zero. For anatase TiO₂ particle, the isoelectric point is around 6.1. This value is considered to be the average of the acid dissociation constants of two kinds of surface hydroxyl groups, *i.e.*, pK_a 2.9 and 9.4. The chemical structure in the acid dissociation state of the two hydroxyl groups on the TiO₂ surface has not been clarified yet but it is considered that the former is the pK_a at which H^+ is attached to O^- of bridged Ti–O–Ti to become $Ti_2 > OH$, and that the latter is the pK_a at which OH^- is detached from terminal Ti–OH to generate Ti^+, or that H^+ binds to TiOH to become $TiOH_2^+$.[9] According to the recent report by calculation, the pH dependence of the surface charge is determined by the combination of the addition of H^+ to $Ti_2 > O^{-0.6}$ with $pK_a = 4.4$ and to $Ti–OH^{-0.38}$ with $pK_a = 7.5$.[10]

As stated in Section 4.2, the equilibrium potential shifts to negative by 59 mV on increase of pH by 1. The fact that pH is 6.1 at zero point potential indicates that in pure water of pH 7.0 the surface of TiO_2 possesses the negative charge of several tens of mV. Actually, as shown in Figure 5.4, negative charge of -25 mV was observed in the low concentration region of NaCl.

On addition of NaCl to such a TiO_2 suspension, Na^+ ions gather on the solid surface to neutralize the surface charge as shown in Figure 5.4. The zeta potential does not change significantly even if the solution contains ethanol up to 50% because water adsorbs more feasibly on the TiO_2 surface than ethanol. For 100% ethanol, the ethoxide adsorbs chemically on the surface, then the number of surface hydroxyl groups decreases and the surface charge becomes nearly zero. On increase of the addition of NaCl, the surface charge becomes positive. This would indicate that instead of ethanol, Na^+ starts to adsorb on the surface.[8]

Hence, TiO_2 surface in the solvents is not necessarily the same as the molecular adsorbed states in vacuum. When the surface charge is large, the electrostatic repulsion among particles takes place and dispersibility in the suspension is enhanced for the powder system. Photocatalytically, the surface charge affects the amount of the adsorption of the reactants, which in turn affects the reaction rates, too.

Under atmospheric conditions, at room temperature, water molecules form several adsorption layers on the surface of TiO_2. The water molecules in the outermost layer are in equilibrium with the water vapor in the air as is observed by ^1H NMR spectroscopy.[11] The dissociation behaviors of a small amount of ethanol on the TiO_2 surface are depicted in Figure 5.5. The loaded ethanol molecules preferably stay in the mobile physisorbed water layer as shown in Figure 5.5(b). On increasing the temperature up to 150 °C along with the evaporation of the mobile water molecules, a small amount of ethanol molecules remains because of the azeotropy. The ethanol molecules reach the solid surface to react with titanol and form ethoxide (c). On reverting the temperature to room temperature, the powder gradually re-adsorbs water molecules in the air to form a physisorbed water layer, and then the ethoxide is hydrolyzed to ethanol, which returns to state (b). Photocatalytic reactions of ethanol under atmospheric conditions take place in these states of the reactant, which decomposes to CO_2 and water (d) *via* several intermediate species.[12]

Thus, the photocatalytic reactions are considered to alter significantly depending on the adsorption states on the TiO_2 surface. Hence,

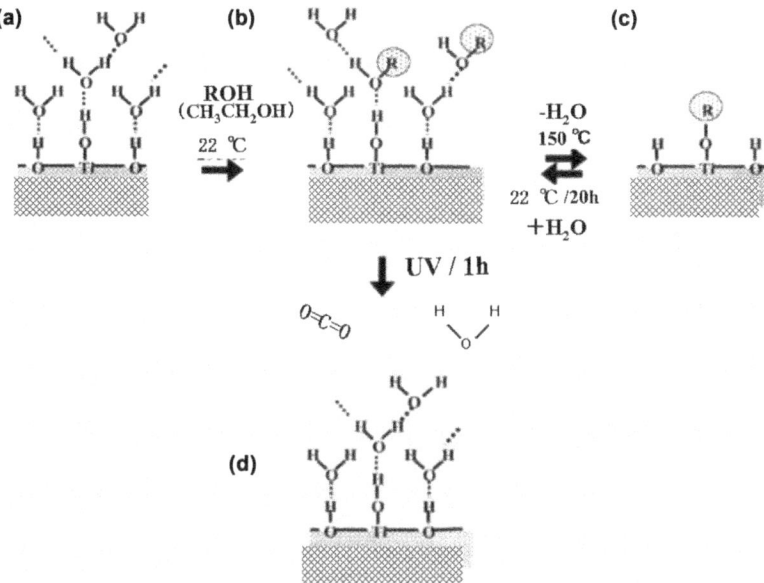

Figure 5.5 Schematic illustration of the incorporation of ethanol into the water layer at the TiO_2 surface and the photocatalytic decomposition in air.
Reprinted with permission from A. Y. Nosaka, Y. Nosaka *et al., Langmuir,* 2003, **19**, 1935.[12] Copyright 2003 American Chemical Society.

it is important to take account of the humidity and the adsorption of reactants to elucidate the reaction mechanisms.

5.2 Trapped Electrons and Holes

On the TiO_2 surface, different from the inner part of the solid, it is highly possible that several energy levels (surface levels) are formed in the bandgap. It is considered that the conduction band electrons and valence band holes generated by the photoabsorption are trapped at each level to be stabilized as trapped electrons and holes, respectively. In the absence of the adsorbed species, the existence of the trapped electrons and holes on the photoirradiated TiO_2 can be confirmed, which cause the oxidation and reduction reactions. To observe such isolated electrons that exist in the materials, electron spin resonance (ESR or EPR) spectroscopy would be a useful method. In Figure 5.6, the representative ESR spectra of the trapped electrons and holes for the commercially available TiO_2 nanoparticles in dried and hydrated states are shown.[13]

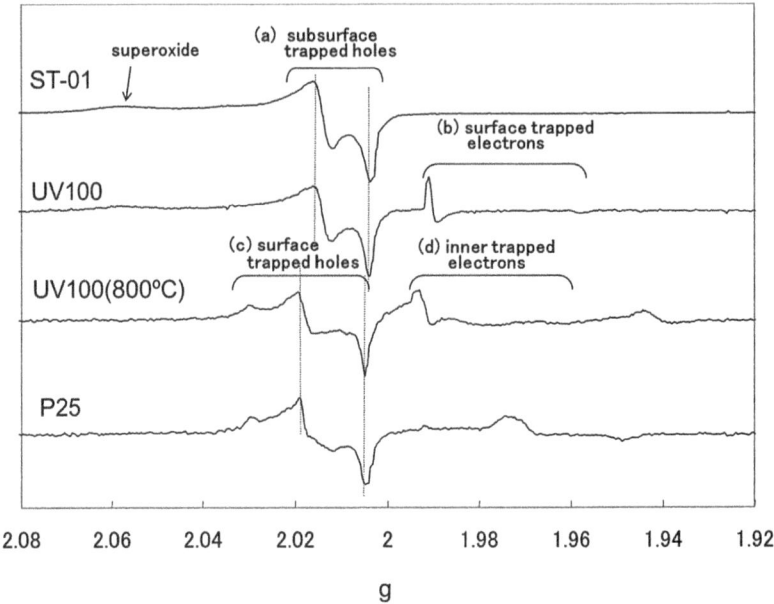

Figure 5.6 ESR spectra with UV irradiation measured under vacuum at 77 K for anatase TiO_2 powders ST-01 (Ishihara Sangyo), Hombikat UV-100 (Sachtleben Chemie), and Degussa P-25 (Nippon Aerosil). Reproduced with permission from T. Hirakawa, Y. Nosaka *et al.*, *J. Phys. Chem. B*, 1999, **103**, 4399. Copyright 1999 American Chemical Society.[13]

For particulate or amorphous TiO_2, the ESR signal for radicals having a distinct structure is depicted by three *g* values. On photo-irradiation, electrons are stabilized at the level just below the conduction band of TiO_2 and become trapped electrons. The ESR signal of the trapped electrons is observed at the higher field $(g < g_e)$ than free electrons $(g_e = 2.0023)$. On the other hand, because trapped holes are generated on the energy level just above the valence band, they are observed at the lower field than free electrons $(g > g_e)$. The five signals in Figure 5.6 correspond to the chemical structures depicted in Figure 5.7, respectively.[14] Namely, in the presence of a large amount of hydroxyl groups, holes are trapped by the subsurface oxygen of TiO_2 possessing OH groups. In the absence of surface hydroxyl groups, the bond of the surface Ti–O–Ti is formed, where holes are trapped. On the other hand, electrons are trapped by the surface Ti atom to form Ti^{3+} or trapped by the inner defection site to form Ti^{3+}. The ESR observation of radicals such as $^\bullet O^-$, $^\bullet O_2^-$, and $HO_2^\bullet$ generated on the surface of TiO_2 nanoparticles has been reported.[15]

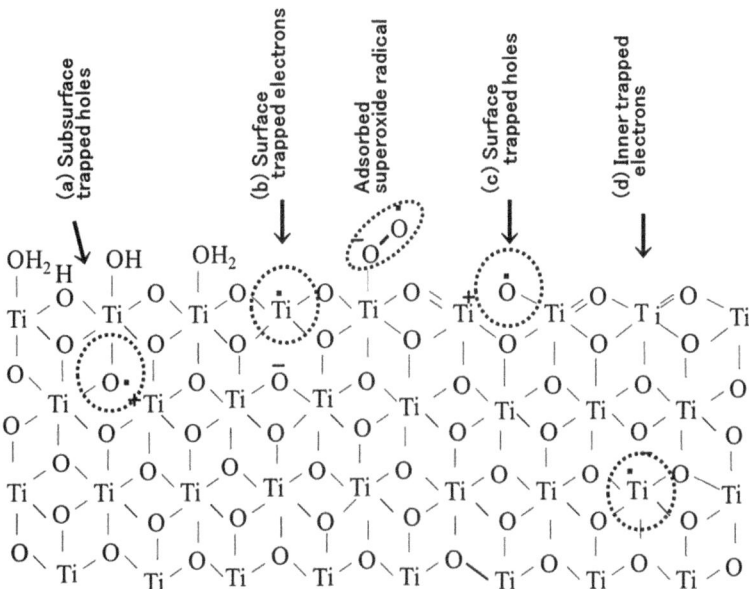

Figure 5.7 Plausible chemical structure for the radicals observed on anatase TiO$_2$ by means of ESR spectroscopy.
Reprinted from Y. Nakaoka and Y. Nosaka, ESR investigation into the effects of heat treatment and crystal structure on radicals produced over irradiated TiO$_2$ powder, *J. Photochem. Photobiol., A*, **110**, 299–305. Copyright (1997), with permission from Elsevier.

The oxidation reaction of the adsorbed molecule would also be caused by the trapped hole because the experiments conducted at low temperature are considered to reflect the reaction steps at room temperature. On ESR observation, the ceased reactions at 77 K begin to proceed on a small increase of temperature, and one can observe the development procedures which would take place at room temperature. Figure 5.8 shows the behaviors of radicals generated on the photoirradiation of the hydrated TiO$_2$ surface and TiO$_2$ surface adsorbed by cysteine.[16] For TiO$_2$ alone, the trapped electrons and holes observed at 77 K diminish at 200 K. On the other hand, for TiO$_2$ on which cysteine molecules are adsorbed, the trapped holes observed at 77 K transferred to cysteine molecules at 150 K and cysteine radicals are formed at 200 K.

ESR is a useful technique with high sensitivity to detect holes and radicals of reaction intermediates but it is not appropriate to investigate the change of the rapid reactions. For the detection of the rapid change of the reaction intermediates, the transient absorption

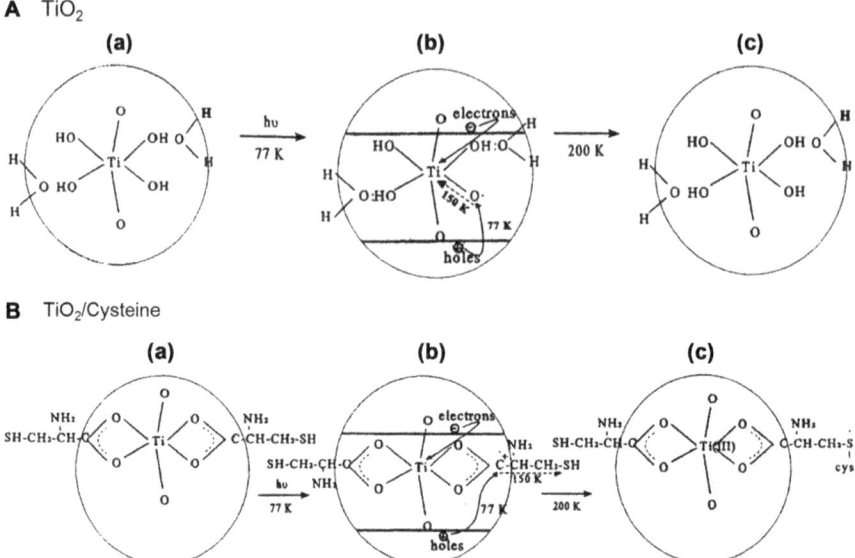

Figure 5.8   Schematic presentation of structure and electron transfer reactions in (A) TiO₂, (B) TiO₂/Cysteine. (a) the structure, (b) and
(c) the electron transfer reactions as detected by ESR spectroscopy following 308 nm irradiation at 77 K.
Reprinted with permission from T. Rajh, M. C. Thurnauer *et al.*,
J. Phys. Chem., 1996, **100**, 4538. Copyright 1996 American
Chemical Society.[16]

spectra from visible to infrared regions are utilized by use of pulsed
lasers.

The trapped electrons and holes could be identified by the transient
spectra as shown in Figure 5.9A,[17] and the assignments of the absorption bands are illustrated in Figure 5.9B. The electrons trapped in
TiO₂ below the conduction band present a broad absorption band at
about 700 nm as illustrated in Figure 5.9A. On the other hand, because the hole-trapping sites locate in the middle of the bandgap, the
absorption peak of the trapped holes at 500 nm ($\sim$2.5 eV) indicates
the transitions from the maximum density of the valence band state
to the trapped state. This absorbance may be also assigned to the
transition from the partially occupied trapped state to the maximum
density of the conduction band state because the bandgap is 3.0–
3.2 eV. Because the holes behave as electron acceptors, the interpretation that the electron in the valence band is excited to the state
of trapped hole seems reasonable. The potential of the hole trapping
site agrees with the report that the luminescence wavelength of the
trapped holes is 810 nm (1.5 eV).[18] The trapped holes are assigned to

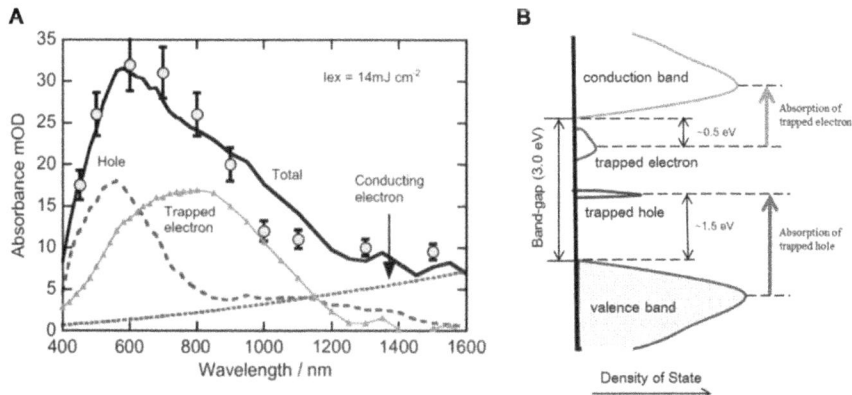

Figure 5.9 (A) Transient absorption spectra of trapped holes, trapped electrons, and conducting electrons. Reprinted from R. Katoh, M. Murai and A. Furube, Transient absorption spectra of nanocrystalline TiO₂ films at high excitation density, *Chem. Phys. Lett.*, **500**, 309–312. Copyright (2010) with permission from Elsevier. (B) Plausible assignments for each absorption band.

Ti–O• formed by the cleavage of the Ti–O–Ti bond in the TiO₂ lattice, and likely decay independently from the electrons because the trapped electron decreases proportionally with the excitation intensity when the number of photons for the excitation is decreased. On the other hand, the trapped hole is not decreased much.[17]

The characteristic absorption that increases with the wavelength without peak observed in Figure 5.9A corresponds to the absorption of free electrons characteristic for semiconductors as stated in Chapter 3 (see eqn (3.5)). This absorption becomes larger with the increase of the oxidation reaction rates which suppresses the recombination with electrons. Furthermore, on increase of the amount of the surface hydration, the free electrons diminish rapidly as the vibration level of Ti–O affects the trapping procedure. The trapped electrons in the shallow trap, which locate at about 0.1 V lower than the conduction band bottom, can be observed by measuring IR absorption spectra under UV irradiation.[19]

Figure 5.10A shows the time profiles of the absorbance for trapped holes measured at 400 nm after 160-fs laser excitation for anatase TiO₂ nanocrystalline film in air, methanol, ethanol, and 2-propanol.[20] In the presence of alcohols, the absorption of the trapped holes decays rapidly, indicating that rapid reactions take place. The reaction rate is the largest for methanol and the lowest for 2-propanol. This order is the reverse for that with OH radicals.

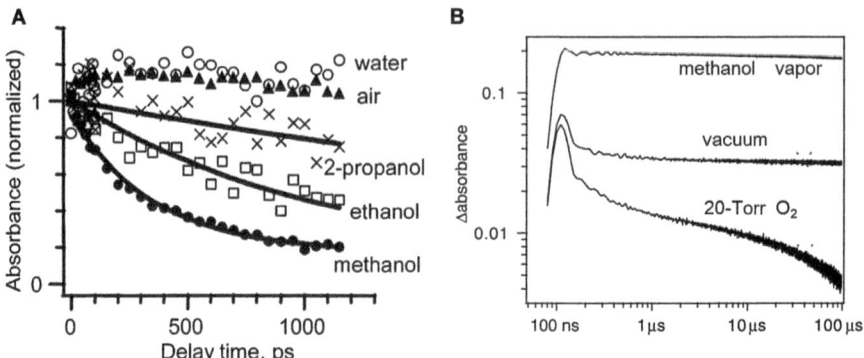

Figure 5.10 (A) Decay of trapped holes measured at 400 nm. Reprinted with permission from ref. 20 (*J. Am. Chem. Soc.*). Copyright 2006 American Chemical Society. (B) Decay of electrons at conduction band measured at $2000\,cm^{-1}$ under various circumstances. Reprinted from A. Yamakata, T. Ishibashi and H. Onishi, Time-resolved infrared absorption study of nine TiO_2 photocatalysts, *Chem. Phys.*, **339**, 133–137. Copyright (2007) with permission from Elsevier.

Figure 5.10B shows the time profiles for conducting electrons in TiO_2 for the transient IR absorption measured at $2000\,cm^{-1}$ (5000 nm) in log scale after the 355 nm pulse excitation in the presence and the absence of 20 Torr O_2. In the presence of oxygen, the electrons at the conduction band diminish to one-third, indicating the reduction of the oxygen.[21] On the other hand, under methanol vapor, which consumes holes, owing to the prevention of the electron–hole recombination, the absorbance of free electrons remains at a higher value.

5.3 Active Oxygen Species

5.3.1 What is Active Oxygen?

The birth of the earth is said to be 4600 million years ago. The first creatures to appear on earth were anaerobic bacteria, which did not need oxygen. Afterwards, cyanobacteria appeared and conducted photosynthesis by use of CO_2 and sunlight, then oxygen was evolved as the waste product. It is believed that the oxygen was accumulated on earth during a long period of time, and finally became the component of the air as today about 2000 million years ago. To protect from the toxicity of oxygen, aerobic creatures which could convert the oxygen to hydrogen peroxide appeared. Then the evolution to today's creatures on earth was initiated.

Because photocatalysts are practically used under aerobic conditions with water vapor, the reactions involving oxygen and water as reaction species are important. Oxygen is a relatively stable molecule which occupies 20% of the air components and the most stable state is usually the triplet state. However, as shown in Figure 5.11, because it possesses an unpaired electron on each of the two degenerate antibonding π-orbitals, the reactivity is not low.

The species with higher reactivity to which oxygen converts are generally called active oxygens. There are four general active oxygens; superoxide anion radical ($^\bullet O_2^-$), singlet oxygen ($^1O_2$), hydroxyl peroxide (H_2O_2), and hydroxyl radical ($^\bullet OH$). Among the active oxygens, $^\bullet O_2^-$, H_2O_2, and $^\bullet OH$ are formed successively by reducing O_2 by one electron, or by oxidizing H_2O by one hole as shown in Figure 5.12. The change of the free energy by reduction and oxidation was shown in the previous chapter (Figure 4.5B).

On the other hand, singlet oxygen is an unstable molecule whose energy is higher than the triplet state oxygen because the combination

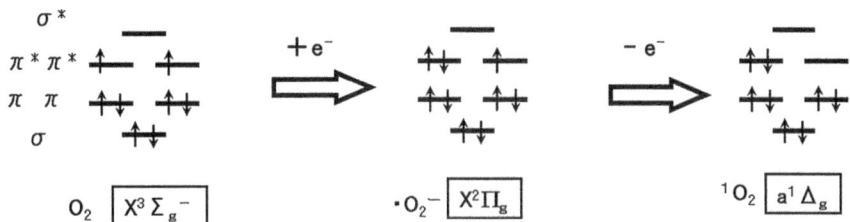

Figure 5.11 The molecular orbitals formed by two p-orbitals of oxygen atoms, showing the ground state of O_2 ($X^3\Sigma_g^-$), superoxide radical, $^\bullet O_2^-$ ($X^2\Pi_g$), and singlet oxygen, $^1O_2$ ($a^1\Delta_g$).

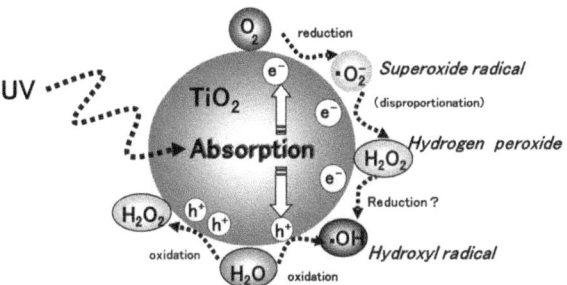

Figure 5.12 Active oxygen species produced in the reduction and oxidation steps of oxygen and water.

of the unpaired election spins is different, although the number of the electrons is the same as that of the oxygen molecule. It is generated by the energy transfer from excited dyes or by the oxidation of $\cdot O_2^-$ as shown in Figure 5.11.[22] In this figure, the symbols surrounded by a square depict the state of electrons on the highest occupied orbital. The initial symbol X expresses the ground state, the small letter a expresses the lowest excited sate with different spin multiplicity, the next superscripts 3, 2, and 1 express the multiplicity of electron spin, Greek symbols Σ, Π, Δ correspond to S, P, and D, which express 0, 1, and 2 of the electron orbital angular momentum. The last subscript g ("gerade" in German) means that the molecular orbital is in the point symmetry and superscript – describes anti-symmetric with respect to a plane through the molecular axis.

Active oxygens alter the chemical structure with pH due to the acid–base equilibrium as shown in Figure 4.5A. Because the addition of a proton to $\cdot O_2^-$ occurs at pH < 4.8, the property is different under neutral and acidic conditions. On the other hand, because the deprotonation takes place at around pH 12 for OH radicals and H_2O_2, the reactions on the corresponding deprotonated ions, *i.e.*, $\cdot O^-$ and HO_2^-, may scarcely occur. There is a review on the detection techniques of all the active oxygens.[23] For individual active oxygens, representative detection methods and the roles in the photocatalytic reactions will be described next.

5.3.2 OH Radicals

Among various active oxygens, the reactivity of the OH radical is considerably higher than the HO_2 radical as will be stated later. Therefore, the OH radical appears often as a reactive species in the reports on photocatalysis. The principal detection methods utilized are as follows:

(a) Direct observation of OH radicals by ESR;
(b) ESR observation of radicals stabilized with a spin trapping reagent;
(c) ESR observation of the decay of nitroxyl radicals by the reaction;
(d) Fluorescence analyses of the reaction products with probe molecules such as terephthalic acid and coumarin;
(e) Color developing method for formaldehyde, which is produced by the reaction with methanol;
(f) Direct measurements of the spectrum of OH radicals by a laser induced fluorescence (LIF) method.

Direct observation of OH radicals by ESR in the gas phase stated in (a) was reported,[24] but it is actually difficult to observe them at room temperature because of the high reactivity.[25] Then, ESR measurements were performed at a low temperature of 77 K aiming at freezing the reaction. For anatase TiO_2, ESR spectra of OH radicals are not observed while the trapped holes are observed. Furthermore, it is confirmed that on increasing temperature from the low temperature, holes diminish due to the electron–hole recombination. As a result water is not oxidized to the OH radical.[26]

The spin trapping method (b) is a conventional method often utilized to detect OH radicals in the reactions in the biological field. This method has been exploited in the photocatalytic researches,[27] where DMPO (5,5-dimethyl-1-pyrroline-*N*-oxide) shown in Figure 5.13A is often utilized as a spin trapping reagent. The unstable OH radicals react with DMPO which is converted to DMPO–OH radicals. They are qualitatively detected by ESR. However, one should keep in mind that DMPO itself can be directly oxidized by photocatalytic reactions.[28] OH radicals produced by the photodecomposition of H_2O_2 can be detected almost completely on the addition of a large amount of DMPO, where the production of DMPO–OH becomes constant. On the other hand, on photoirradiating TiO_2, the produced amount of DMPO–OH radical is not saturated even with further increase of the amount of DMPO added. This indicates the possibility that DMPO does not trap the OH radical generated by the photocatalytic reactions but that DMPO itself produces DMPO–OH radicals. Furthermore, DMPO can be also oxidized by the photosensitized reactions because it absorbs ultraviolet light though slightly as will be described in the next chapter. Therefore, one should be cautious to conclude that the OH radicals are involved in photocatalytic reactions on the basis of the experimental results with DMPO.

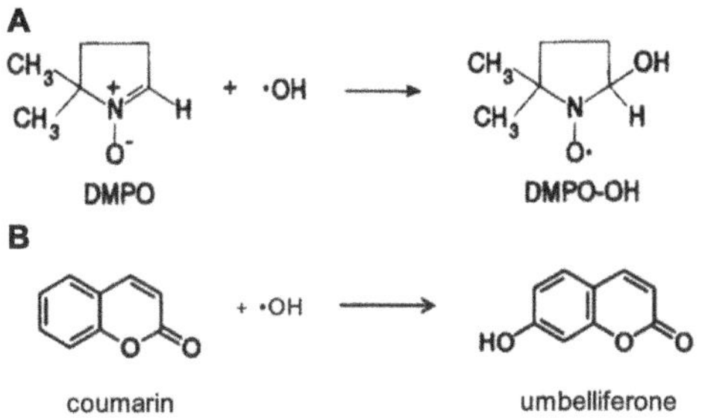

Figure 5.13 Reactions for detecting OH radicals with (A) a spin trapping reagent, DMPO, and (B) a fluorescence probe reagent, coumarin.

One should take care over the quantitative detection of OH radicals by measuring the decay of the stable nitroxyl radicals[29] stated above (c) because the stable nitroxyl radical itself can be oxidized and/or reduced at the electrode.[30] Thus, oxidation by the holes at the valence band or the trapped holes is sometimes incorrectly regarded as oxidation by OH radicals.[31]

The method (d) to detect the fluorescence is a technique often utilized in the field of radiation chemistry. With this method fluorescent products are generated by the selective reactions of terephthalic acid[32] and coumarin[33] with OH radicals as shown in Figure 5.13B. From the fluorescence intensity, the amount of OH radicals in TiO_2 photocatalytic systems was qualitatively estimated for the first time.[34] Actually, 45% of the produced OH radicals can be detected quantitatively with the use of terephthalic acids[35] while 6% of those can be detected with the use of coumarin.[36] On adopting terephthalic acid, it is effective in alkaline solutions and the amount of the products are saturated at the high concentration of terephthalic acid. However, it is not oxidized by OH radicals at a pH below 4 but it is oxidized by holes with the yield of 8%.[37] On the other hand, coumarin self-decomposes gradually in alkaline solutions. The quantum yield of OH radicals detected for TiO_2 photocatalytic systems by the use of these fluorescent probe reagents is extremely small, on the order of 10^{-5} to 10^{-4}.[35,36]

The method (e) is based on the reaction that the OH radical reacts with methanol to produce formaldehyde. The quantum yield of the formation of OH radicals obtained with a color developing method is reported as 4%.[38] On the basis of the results of ESR measurements at low temperature, methanol is directly oxidized by trapped holes without the contribution of OH radicals.[39] At room temperature in solution, methanol can also be oxidized on the surface of TiO_2 without the contribution of OH radicals.[40]

The LIF method (f) is the method of high sensitivity that is utilized to detect a very small amount of OH radicals in the atmosphere. Because the fluorescence due to OH radicals appears at 310 nm, the intensity is measured as a function of the excitation wavelength by using a dye laser as shown in Figure 5.14A. It was confirmed that OH radicals are dispersed from the photoexcited TiO_2 surface, at which the OH radicals are considered to be generated not by the oxidation of water but from the hydrogen peroxide on the TiO_2 surface (Figure 5.14B).[41] This method is one of a few methods with which OH radicals can be directly observed. However, because the fluorescence of OH radicals is quenched with water vapor, the observations are

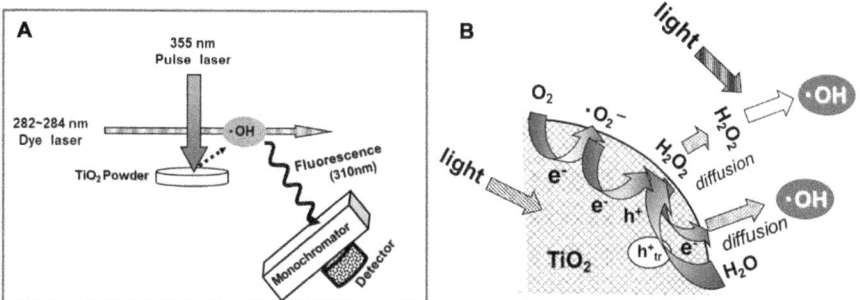

Figure 5.14 (A) Laser-induced-fluorescence detection of OH radicals released from the irradiated TiO₂ surface. (B) The plausible mechanism of OH radical formation.[41]
Reprinted from Y. Nosaka, *Environmentally Benign Photocatalysts – Applications of Titanium Oxide-based Materials*, Surface Chemistry of TiO₂ Photocatalysis and LIF Detection of OH Radicals, 2010, ch. 8, p. 206 and p. 212, with permission from Springer Science + Business Media.

limited in the gas phase and this method cannot be absolutely used in aqueous solution.

Thus, there are many methods with which OH radicals can be detected. One should be very cautious about the conditions on the applications as well as the interpretation of the results. The contribution of OH radicals to the reaction mechanism will be described in the next chapter (Section 6.3).

5.3.3 Superoxide Radical

Because photocatalysts are often utilized under atmospheric conditions, the photoinduced CB electrons transfer to the oxygen in the air to generate $^{\bullet}O_2^{-}$. Although the photoabsorption of $^{\bullet}O_2^{-}$ is observed at 250 nm,[42] it is difficult to measure directly with the absorbance due to the absorption of photocatalysts. Then the methods that can be applied would be as follows:

(a) Direct observation of $^{\bullet}O_2^{-}$ by ESR by freezing at low temperature;
(b) Detection with an ESR spin trapping method using the trapping reagents such as DMPO;
(c) The color reaction with NBT (Nitro Blue Tetrazolium);
(d) Chemiluminescence method with luminol or luciferin analogs (MCLA).

With ESR measurements by freezing at low temperature (a), the coordinated state of $^{\bullet}O_2^-$ can be observed. Because the free states of $^{\bullet}O_2^-$ cannot be observed by ESR due to the degeneration of the spin orbit,[25] it is necessary to resolve the degeneracy by coordinating to Ti^{4+}. The broad signal at lower field ($g = 2.06$) in Figure 5.6 is considered to be a part of the signals of $^{\bullet}O_2^-$ adsorbed on the surface.[14] The molecular oxygen adsorbs on the oxygen defects or 5-coordinate Ti^{4+} on the surface. Therefore, the location of the adsorbed oxygen depicted in Figure 5.7 is not necessarily the rigid one. The signal does not appear like that at the lower field (left side) but it often presents spectra similar to those of the trapped hole depending on the strength of the adsorption.[43] Namely, it is suspected that the signal designated as superoxide in Figure 5.6 corresponds to that of $^{\bullet}O_2^-$ weakly bound to the surface because the stronger the adsorption is, the peak at lower field shifts to the higher field. Superoxide radicals in the form of $^{\bullet}O_2^-$ strongly bound to the surface (Ti^{4+}–$OO^{\bullet}$) may present spectra similar to those of trapped holes.

As for the ESR detection of $^{\bullet}O_2^-$ utilizing trapping reagent DMPO (b), it is known that the $^{\bullet}O_2^-$ adduct of DMPO is unstable and converts to the OH radical adduct.[44] For the NBT method (c), the yellow NBT becomes blue formazan due to the reduction by $^{\bullet}O_2^-$. The low solubility of the products and the selectivity of the reactions may be the problem.

For the chemiluminescence method with luminol and luciferin analogs (MCLA) (d), as shown in Figure 5.15A and B, the excited molecules are generated on the reactions. Therefore, this is a highly sensitive method that enables the detection of a very small amount of the products. On adopting luminol, as shown in Figure 5.15A, the reaction proceeds as follows: in alkaline solution luminol takes one-electron oxidation by $^{\bullet}O_2^-$ and then reacts with $^{\bullet}O_2^-$ to generate the excited state.[45] On the other hand, on adopting MCLA the reactions proceed directly.[46]

For anatase TiO_2 films, $^{\bullet}O_2^-$ is generated under very weak photo-irradiation such as 1 $\mu W\,cm^{-2}$ with the density of $1 \times 10^{14}\,cm^{-2}$ under atmospheric conditions, and $2 \times 10^{14}\,cm^{-2}$ in aqueous solution, respectively. The quantum yields are reported to be 0.4 and 0.8, respectively, which are significantly high.[47]

The lifetime of $^{\bullet}O_2^-$ is the longest next to that of H_2O_2 among the active oxygen species and it becomes H_2O_2 by the following disproportionation reaction:

$$^{\bullet}O_2^- + HO_2^{\bullet} + H_2O \rightarrow H_2O_2 + O_2 + OH^- \qquad (5.1)$$

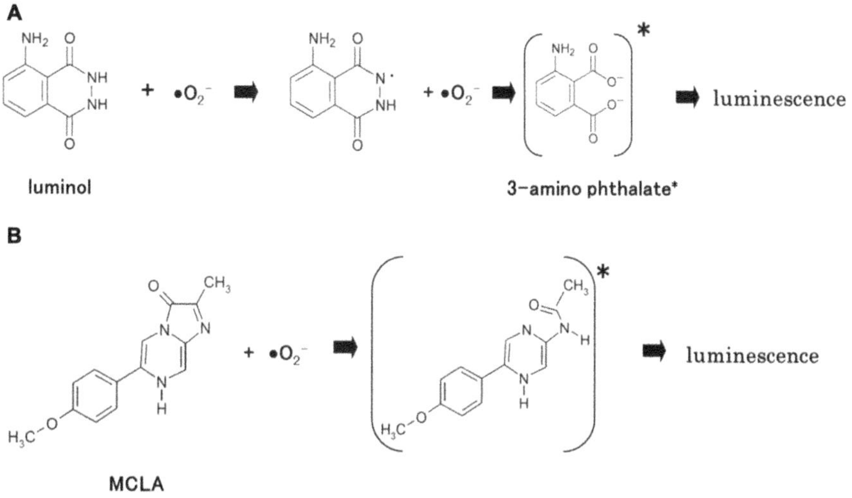

Figure 5.15 Reactions for detecting $\cdot O_2^-$ with (A) luminol and (B) MCLA by chemiluminescence.

Because $\cdot O_2^-$ takes the form of peroxyl radical of $HO_2^\cdot$ below pH 4.8, the bimolecular rate constant for the reaction (5.1) becomes largest at around pH 5. The observed bimolecular rate constant k_{obs} becomes smaller with an increase of pH and is expressed as $k_{obs} = 6 \times 10^{12-pH}$ M^{-1} s^{-1}.[42] Therefore, the measurement of $\cdot O_2^-$ is not difficult, especially in alkaline solution.

The decay process of $\cdot O_2^-$ in the TiO_2 powder suspension after the photoirradiation is shown in Figure 5.16A. On the analysis of this decay curve, it turns out that it is not single exponential decay nor second-order decay but it is described as the following simple equation.[48]

$$[\cdot O_2^-] = [\cdot O_2^-]_0 \times t^{\beta-1} \tag{5.2}$$

This type of reaction is called fractal like kinetics. Eqn (5.2) does not contain an exponential function of time as is generally expressed as time dependency. Though time t usually appears in the form of e^{-kt} in the rate equation, in this case the rate constant k is extremely small so that this term is considered to become 1. When $\cdot O_2^-$ is adsorbed on the surface where the surface trapped holes are placed nearby, it is oxidized by the trapped holes depending on the distance between the hole and $\cdot O_2^-$. This means that the time dependency on the two species is determined by the statistical distribution of the distance

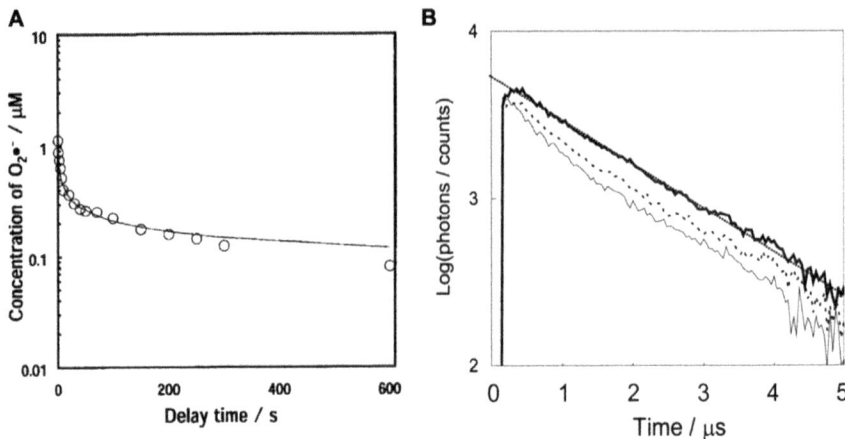

Figure 5.16   (A) Time dependence of the $^{\bullet}O_2^{-}$ concentration after stopping the irradiation on the TiO_2 suspension. The solid line shows the curve fitted by eqn (5.2). Reprinted with permission from H. Hirakawa and Y. Nosaka, *Langmuir*, 2002, **18**, 3247. Copyright 2002 American Chemical Society.[48] (B) Time dependence of phosphorescence intensity of $^{1}O_2$ monitored at 1250 nm after the pulsed excitation on P25 TiO_2 aqueous suspension. Without additives (heavy line), with methionine (dashed line), and folic acid (fine line). Reprinted from T. Daimon, T. Hirakawa and Y. Nosaka, *Electrochemistry*, 2008, **76**, 136, with permission from Electrochemical Society of Japan.[55]

between $^{\bullet}O_2^{-}$ and the trapped holes which distribute randomly on the surface. Furthermore, β is a parameter relevant to the uniformity of the reaction and when it becomes closer to 1, one can regard that the oxidation of $^{\bullet}O_2^{-}$ takes place on the relatively uniform surface.

Generally the reactivity of $^{\bullet}O_2^{-}$ is low although the reaction with molecules with high electron affinity is rapid. The reactivity of $HO_2^{\bullet}$ seems high compared to that of OH radicals as is suspected from the molecular form. In Figure 5.17A and B, in order to take the correlations the second-order rate constant of $HO_2^{\bullet}$ is plotted against the second-order rate constants of $^{\bullet}O_2^{-}$ and $^{\bullet}OH$. The second-order rate constants in the figure are reported for organic and inorganic compounds, metal complexes and amino acids.[42,49] As shown in Figure 5.17A, the difference of the reactivity between $^{\bullet}O_2^{-}$ and $HO_2^{\bullet}$ is not significantly large. But as shown in Figure 5.17B it is notable that the reactivity of the OH radical is the highest among all the compounds adopted here and the reaction rate constant is nearly the rate of diffusion controlled reaction.

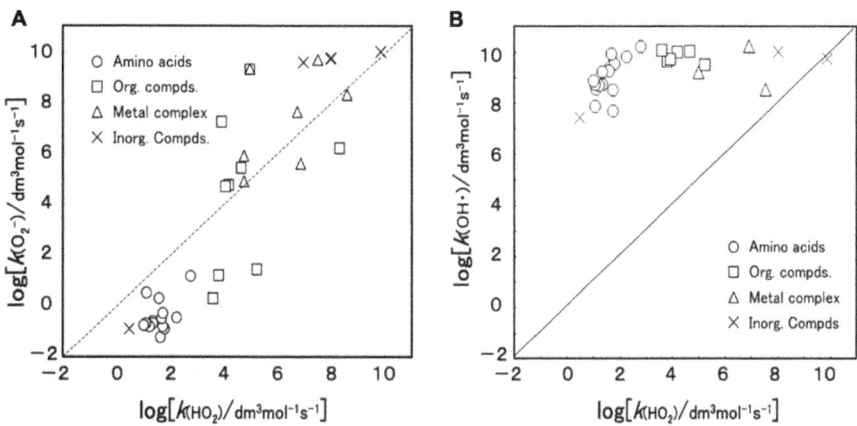

Figure 5.17 Relationships between bi-molecular rate constants of (A) $^{\bullet}O_2^-$ to $HO_2^{\bullet}$ and (B) $^{\bullet}OH$ to $HO_2^{\bullet}$ for the reactions with various molecules taken from ref. 42 and 49.

5.3.4 Singlet Oxygen

Because the singlet oxygen $^1O_2$ is an excited state of O_2, it is de-activated to the original stable O_2 without being involved in chemical reactions. As detection methods of $^1O_2$ the following techniques are known:

(a) Direct observation of the light emission by $^1O_2$;
(b) By using chemiluminescence reagents for $^1O_2$, light emission is observed;
(c) Analysis of the products generated by the reaction with furans or other molecules;
(d) ESR detection of radicals generated by the reaction with probe reagents such as 2,2,6,6-tetramethyl piperidine.

Two kinds of emission are known for the direct observation of $^1O_2$ for (a). The dimol emission $(2\ ^1\Delta_g \rightarrow 2\ ^3\Sigma_g^-)$ is at 633 nm, which is in the visible light region. Therefore, the detection method is simple but it is difficult to discriminate from the other emission. On the other hand, the observation of the phosphorescence $(^1\Delta_g \rightarrow ^3\Sigma_g)$ of 1270 nm is the most reliable.[50]

It was reported in the past that the chemiluminescence could be detected by the use of luminol and MCLA for (b). However, the chemiluminescence has been ascribed to that of $^{\bullet}O_2^-$ but not to $^1O_2$, and novel reagents specific to $^1O_2$ have been developed.[51] For the reaction (c) diphenyl furan and cholesterol are used. However, because these molecules are insoluble in water, they are used by adsorbing on the particles or putting the micelle in aqueous solutions. The method (d) to produce nitroxide radicals has been used for a long time. Because the same radical product is formed by the reaction of valence band holes with $^{\bullet}O_2^{-}$ in addition to $^1O_2$, one should be cautious to apply this to photocatalysis.[52]

The fact that singlet oxygen is formed for the photocatalytic systems of TiO_2 has been confirmed by detecting the phosphorescence of the $^1\Delta_g$ state.[53] The procedure of $^1O_2$ generation would be the oxidation of $^{\bullet}O_2^{-}$ as shown in Figure 5.11. Because three electrons in the antibonding π-orbitals of $^{\bullet}O_2^{-}$ are generated on the reduction of O_2, there are three plausible ways for the electrons to be removed on the oxidation, which form the different oxygen molecules in the different electronic states.[22] Namely, those are the three in the order of the energy state, $^3\Sigma_g^{-}$, $^1\Delta_g$, and $^1\Sigma_g^{+}$. The lifetime of $^1\Sigma_g^{+}$ with high energy is extremely short and it transfers immediately to the $^1\Delta_g$ state.[50] Therefore, $^1O_2(^1\Delta_g)$ is considered to be formed with the probability of 2/5.

The lifetime of $^1\Delta_g$ is several tens of milliseconds in the atmosphere, while in water it becomes as short as several microseconds as shown in Figure 5.16B. Because the lifetime is determined by the process of the energy transfer to the vibrational energy of the solvent, it becomes longer in deuterium oxide as 60 μs and even longer in organic solvent. Singlet oxygen is said to present high reactivity in some organic molecules such as olefin and amine.[54] However, because it exists at the electronically exited state, the reactivity might be lost without reacting. Actually, as shown in Figure 5.16B, the decay of $^1O_2$ does not change significantly on the addition of reactive molecules, such as methionine and folic acid.[55]

5.3.5 Hydrogen Peroxide

Hydrogen peroxide (H_2O_2) is a stable compound formed by the reduction of $^{\bullet}O_2^{-}$ as shown in Figure 5.12 or by the disproportionation reaction of $^{\bullet}O_2^{-}$ indicated in eqn (5.1). In the case of photocatalytic oxidation, generation of the OH radical is rare but water can be oxidized by two holes to produce H_2O_2 on the surface of the photocatalyst.[56] Because H_2O_2 is photocatalytically formed *via* several reaction procedures, the amount of the production largely depends on the reaction conditions.

The observation methods of H_2O_2 are known as follows:

(a) Iodometry;[57]
(b) Fluorescence probe method;[58]
(c) Chemiluminescence method.

As chemiluminescence methods (c), the method with luminol which is oxidized by hemoglobin (Figure 5.18A)[59] and that with lucigenin (Figure 5.18B)[60] are used for photocatalytic systems in aqueous solution. For both methods, because H_2O_2 is strongly adsorbed on the surface of the metal oxides, quantitative measurement of the accurate amount of the production is considered difficult.

It is known that as shown in Figure 5.12, H_2O_2 is formed by the reduction of O_2 but that it reverts to H_2O and O_2 by disproportionation on the various surfaces of the photocatalysts.

$$2H_2O_2 \rightarrow O_2 + 2H_2O \tag{5.3}$$

As shown in Figure 5.12, OH radicals may be formed on further reduction of H_2O_2.

$$H_2O_2 + e^- \rightarrow {}^\bullet OH + OH^- \tag{5.4}$$

However, because the $^\bullet OH$ radical is unstable as shown in Chapter 4 (Figure 4.5B), one-electron reduction is not enough; H_2O_2

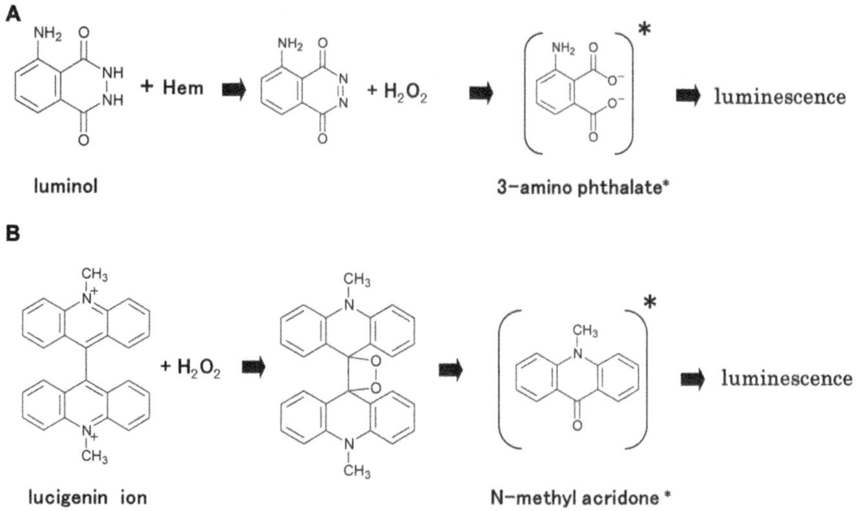

Figure 5.18 Chemiluminescence reactions for detection of H_2O_2 using (A) luminol with hemoglobin (Hem), and (B) lucigenin.

takes two-electron reduction to become water.[36] Even if $^\bullet OH$ is formed, it reacts with H_2O_2 by the rapid reaction $(k = 3 \times 10^7 \ M^{-1} \ s^{-1})$ to form $^\bullet O_2^-$, eqn (5.5).

$$H_2O_2 + {}^\bullet OH \rightarrow {}^\bullet O_2^- + H_2O + H^+ \tag{5.5}$$

The reaction (5.6) in which OH radicals are formed by the reaction of H_2O_2 with $^\bullet O_2^-$, which is known as the Haber–Weiss reaction, may be possible.

$$H_2O_2 + {}^\bullet O_2^- \rightarrow {}^\bullet OH + O_2 + OH^- \tag{5.6}$$

However, this reaction is actually negligible for photocatalytic systems, because the generated amount of H_2O_2 and $^\bullet O_2^-$ are low and the bimolecular reaction rate constant is small $(k = 0.13 \ M^{-1} \ s^{-1})$.[42] In the presence of the abundant H_2O_2, the OH radical becomes $^\bullet O_2^-$ as indicated in eqn (5.5).

References

1. U. Diebold, *Surf. Sci. Rep.*, 2003, **48**, 53.
2. M. A. Henderson, *Surf. Sci. Rep.*, 2011, **66**, 185.
3. M. A. Henderson and I. Lyubinetsky, *Chem. Rev.*, 2013, **113**, 4428.
4. M. A. Henderson, *Langmuir*, 1996, **12**, 5093.
5. G. D. Parfitt, *Prog. Surf. Membr. Sci.*, 1976, **11**, 181.
6. Y. Suda and T. Morimoto, *Langmuir*, 1987, **3**, 786.
7. K. I. Hadjiivanov and D. G. Klissurski, *Chem. Soc. Rev.*, 1996, **25**, 61.
8. W. Janusz, A. Sworska and J. Szczypa, *Colloids Surf., A*, 1999, **152**, 223.
9. P. A. Connor, K. D. Dobson and A. S. McQuillan, *Langmuir*, 1999, **15**, 2404.
10. K. Bourikas, C. Kordulis and A. Lycourghiots, *Chem. Rev.*, 2014, **114**, 9754.
11. A. Y. Nosaka and Y. Nosaka, *Bull. Chem. Soc. Jpn.*, 2005, **78**, 1595.
12. A. Y. Nosaka, Y. Nosaka *et al.*, *Langmuir*, 2003, **19**, 1935.
13. T. Hirakawa, Y. Nosaka *et al.*, *J. Phys. Chem. B*, 1999, **103**, 4399.
14. Y. Nakaoka and Y. Nosaka, *J. Photochem. Photobiol., A*, 1997, **110**, 299.
15. J. M. Coronado *et al.*, *Langmuir*, 2001, **17**, 5368.
16. T. Rajh, M. C. Thurnauer *et al.*, *J. Phys. Chem.*, 1996, **100**, 4538.

17. R. Katoh, M. Murai and A. Furube, *Chem. Phys. Lett.*, 2010, **500**, 309.
18. R. Nakamura, Y. Nakato *et al.*, *J. Am. Chem. Soc.*, 2005, **127**, 12975.
19. D. M. Savory and A. J. McQuillan, *J. Phys. Chem. C*, 2013, **117**, 23645.
20. Y. Tamaki, A. Furube *et al.*, *J. Am. Chem. Soc.*, 2006, **128**, 416.
21. A. Yamakata, T. Ishibashi and H. Onishi, *Chem. Phys.*, 2007, **339**, 133.
22. H. Saito and Y. Nosaka, *J. Phys. Chem. C*, 2014, **118**, 15656.
23. P. Fernandez-Castro, I. Ortiz *et al.*, *J. Chem. Technol. Biotechnol.*, 2015, **90**, 796.
24. J. M. Brown, M. Kaiseab, C. M. L. Kerra and D. J. Milton, *Mol. Phys.*, 1977, **36**, 553.
25. J. A. Weil and J. R. Bolton, *Electron Paramagnetic Resonance*, John-Wiley Pub., 2007.
26. O. I. Micic, M. C. Thurnauer *et al.*, *J. Phys. Chem.*, 1993, **97**, 7277.
27. C. D. Jaeger and A. J. Bard, *J. Phys. Chem.*, 1979, **83**, 3152.
28. M. A. Grela, M. E. J. Coronel and A. J. Colussi, *J. Phys. Chem.*, 1996, **100**, 16940.
29. P. F. Schwarz *et al.*, *J. Phys. Chem. B*, 1997, **101**, 7127.
30. S. Kishioka *et al.*, *Electrochem. Acta*, 2003, **48**, 1589.
31. Y. Nosaka *et al.*, *Phys. Chem. Chem. Phys.*, 2003, **5**, 4731.
32. R. W. Matthews, *Radiat. Res.*, 1980, **83**, 27.
33. G. Louit, S. Pin *et al.*, *Radiat. Phys. Chem.*, 2005, **72**, 119.
34. K. Ishibashi, T. Ohsaka and K. Tokuda, *J. Photochem. Photobiol., A*, 2000, **134**, 139.
35. Y. Nakabayashi and Y. Nosaka, *Phys. Chem. Chem. Phys.*, 2015, **17**, 30570.
36. J. Zhang and Y. Nosaka, *J. Phys. Chem. C*, 2014, **118**, 10824.
37. A. V. Taborda, M. A. Brusa and M. A. Grela, *Appl. Catal., A*, 2001, **208**, 419.
38. L. Sun and J. R. Bolton, *J. Phys. Chem.*, 1996, **100**, 4127.
39. O. I. Micic, M. C. Thurnauer *et al.*, *J. Phys. Chem.*, 1993, **97**, 13284.
40. J. Zhang and Y. Nosaka, *Appl. Catal., B*, 2015, **166**, 32.
41. Y. Murakami, Y. Nosaka *et al.*, *J. Phys. Chem. C*, 2007, **111**, 11339.
42. B. H. J. Bielski *et al.*, *J. Phys. Chem. Ref. Data*, 1985, **14**, 1041.
43. M. Che and A. J. Tench, *Adv. Catal.*, 1983, **32**, 1.
44. H. Noda *et al.*, *Bull. Chem. Soc. Jpn.*, 1994, **67**, 2031.
45. Y. Koizumi and Y. Nosaka, *J. Phys. Chem. A*, 2013, **117**, 7705.
46. N. Suzuki *et al.*, *Agric. Biol. Chem.*, 1991, **55**, 157.
47. K. Ishibashi, K. Hashimoto *et al.*, *J. Phys. Chem. B*, 2000, **104**, 4934.

48. H. Hirakawa and Y. Nosaka, *Langmuir*, 2002, **18**, 3247.
49. G. V. Buxton *et al.*, *J. Phys. Chem. Ref. Data*, 1988, **17**, 513.
50. C. Schweitzer and R. Schmidt, *Chem. Rev.*, 2003, **103**, 1685.
51. H. Wu, Q. Song *et al.*, *Trends Anal. Chem.*, 2011, **30**, 133.
52. Y. Nosaka *et al.*, *J. Phys. Chem. B*, 2006, **110**, 12993.
53. Y. Nosaka *et al.*, *Phys. Chem. Chem. Phys.*, 2004, **6**, 2917.
54. F. Wilkinson, W. P. Helman and A. B. Rosss, *J. Phys. Chem. Ref. Data*, 1995, **24**, 663.
55. T. Daimon, T. Hirakawa and Y. Nosaka, *Electrochemistry*, 2008, **76**, 136.
56. Y. Kakuma, A. Y. Nosaka and Y. Nosaka, *Phys. Chem. Chem. Phys.*, 2015, **17**, 18691.
57. R.-A. Doong and W.-H. Chang, *J. Photochem. Photobiol., A*, 1997, **107**, 239.
58. K. Ishibashi, K. Hashimoto *et al.*, *Electrochemistry*, 2001, **69**, 160.
59. T. Hirakawa and Y. Nosaka, *J. Phys. Chem. C*, 2008, **112**, 15818.
60. J. Oguma, Y. Nosaka *et al.*, *Appl. Catal., B*, 2013, **129**, 282.

6 Kinetics and Mechanism in Photocatalysis

6.1 Time Scales of Photocatalysis

The time scale of photocatalysis is summarized in Figure 6.1. Because holes and electrons are generated simultaneously on photoabsorption, they are considered to be generated within 1 fs of the period of the electromagnetic wave. The rates trapped by TiO_2 can be measured by the transient absorbance, as was shown in Chapter 5 (Figure 5.9). The trapping rate of the holes is shorter than that of the electrons. The energy level of the trapped electron locates 0.1–0.5 eV below the conduction band and the electrons in the shallow trap are considered in equilibrium with the free electrons for a while.[1] Because the level of one-electron reduction of O_2 is close to the edge of the conduction band (see Figure 4.5A), in the presence of oxygen the reduction of O_2 would be caused by the electron transfer from the conduction band or from the state of shallow trap. The reaction rate constant is reported[2] as 7.6×10^7 M^{-1} s^{-1} and the lifetime is relatively long at 10 μs because the usual oxygen concentration dissolved in water is around 1 mM. The generated $\bullet O_2^-$ is further reduced or produces H_2O_2 by the disproportionation reaction with $HO_2 \bullet$ as was described in Chapter 5 (Section 5.3.3). The recombination procedure reportedly takes place within 20 ns although the recombination rate of the electrons with the holes likely varies depending on the characteristics of photocatalysts.[3] The recombination competes with the oxidation of reactants to form radicals which react with oxygen in scales from μs to ms. Details can be found in the literature.[4]

Introduction to Photocatalysis: From Basic Science to Applications
By Yoshio Nosaka and Atsuko Nosaka
© Yoshio Nosaka and Atsuko Nosaka, 2016
Published by the Royal Society of Chemistry, www.rsc.org

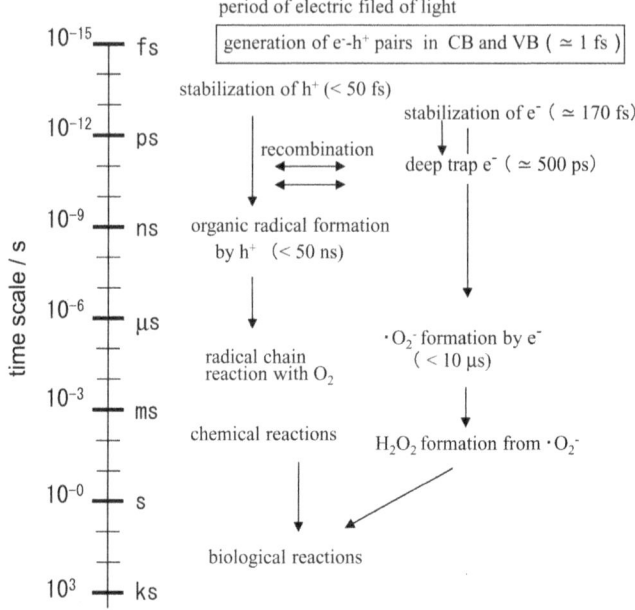

Figure 6.1 Time scale of photocatalysis with TiO$_2$.

6.2 Basic Kinetic Analysis

6.2.1 Modelling of Photocatalytic Processes

To examine the reaction efficiency and the effect of light intensity, a simple reaction model will be introduced though the actual photocatalytic reactions proceed *via* several complicated steps. As a simple reaction model which describes the whole process regarding the efficiency and the reaction rate for photocatalytic reactions, 11 processes shown in Figure 6.2 can be considered. In the figure, A and D represent the molecules which accept and donate electrons, *i.e.*, molecules to be reduced and oxidized, respectively.

(a) Photo-excitation process
   ① Conduction band electrons (e$^-_{CB}$) and valence band holes (h$^+_{VB}$) are generated by the absorption of light by the photocatalyst. The rate of this process corresponds to that of the absorption of the light.
(b) Reduction process
   ② Conduction band electrons reduce the adsorbed molecules.

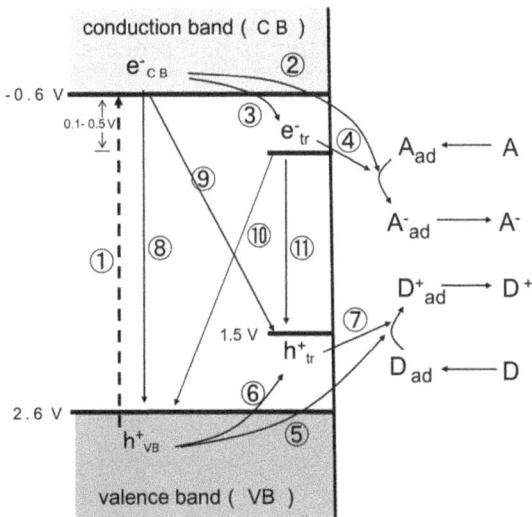

Figure 6.2 General reaction paths of photocatalysis with the oxidation of molecule D and the reduction of molecule A.

③ Conduction band electrons are trapped on the surface of the photocatalysts to be trapped electrons (e^-_{tr}).

④ The trapped electrons reduce the adsorbed molecules.

(c) Oxidation process

⑤ Valence band holes oxidize the adsorbed molecules.

⑥ Valence band holes are trapped on the surface of the photocatalyst to become trapped holes (h^+_{tr}).

⑦ The trapped holes oxidize the adsorbed molecules.

(d) Recombination process

⑧ Conduction band electrons recombine with the valence band holes.

⑨ Conduction band electrons recombine with the trapped holes.

⑩ Trapped electrons recombine with the valence band holes.

⑪ Trapped electrons recombine with the trapped holes.

Besides the thermal relaxation process, the light emission process by which the energy converts to the photon energy may be involved in every recombination process. However, the contributions of most of the photoemission processes are very slight.

Among the above processes, there is a case in which the reaction takes place during the trapping process of holes and electrons. Therefore, to make the discussion simple, (b) reduction process

② ~ ④, (c) oxidation process ⑤ ~ ⑦, and (d) recombination process ⑧ ~ ⑪ are expressed all together as one process as follows:

(a) Photo-excitation process (①)

$$TiO_2 \xrightarrow{g} e^- + h^+ \tag{6.1}$$

(b) Reduction process (② ~ ④)

$$e^- + A_{ad} \xrightarrow{k_e} A^- \tag{6.2}$$

(c) Oxidation process (⑤ ~ ⑦)

$$h^+ + D_{ad} \xrightarrow{k_h} D^+ \tag{6.3}$$

(d) Recombination process (⑧ ~ ⑪)

$$e^- + h^+ \xrightarrow{k_r} TiO_2 \tag{6.4}$$

where the generation rate of the electron-hole pair g is the same as the light absorption rate and expressed with αI as described in Section 2.1, where α is the absorption coefficient (cm^{-1}), and I is the intensity of the light irradiation ($mol\,cm^{-2}\,s^{-1}$). The reaction equations for electrons and holes are expressed as follows based on eqn (6.1)–(6.4).

$$\frac{d[e^-]}{dt} = \alpha I - k_e[e^-][A_{ad}] - k_r[e^-][h^+] \tag{6.5}$$

$$\frac{d[h^+]}{dt} = \alpha I - k_h[h^+][D_{ad}] - k_r[e^-][h^+] \tag{6.6}$$

On the other hand, the rate equations with which the adsorbed molecules are reduced and oxidized are expressed as eqn (6.7) and (6.8), respectively.

$$\frac{d[A^-]}{dt} = k_e[e^-][A_{ad}] \tag{6.7}$$

$$\frac{d[D^+]}{dt} = k_h[h^+][D_{ad}] \tag{6.8}$$

6.2.2 Quantum Yield

Firstly, take a glance at the quantum yield of the oxidation reaction. The quantum yield ϕ is the ratio of the number of the product molecules to the number of the photons absorbed, namely, the ratio of the generation rate of oxidized product D^+ to the generation rate of the electron-hole pairs. Therefore, the quantum yield of D^+ can be expressed as follows.

$$\phi = \frac{\dfrac{d[D^+]}{dt}}{g} \tag{6.9}$$

ϕ can be expressed as eqn (6.10) from eqn (6.8),

$$\phi = \frac{k_h[h^+][D_{ad}]}{\alpha I} \tag{6.10}$$

On applying steady state approximation for $[e^-]$ and $[h^+]$, by setting 0 for eqn (6.5) and (6.6), the following relationship can be obtained.

$$\alpha I - k_h[h^+][D_{ad}] - k_r[e^-][h^+] = 0 \tag{6.11}$$

$$k_e[e^-][A_{ad}] = k_h[h^+][D_{ad}] \tag{6.12}$$

From eqn (6.10)–(6.12), the following relationship is derived.

$$\frac{1-\phi}{\phi^2} = \frac{k_r}{k_e k_h[A_{ad}][D_{ad}]}\alpha I \tag{6.13}$$

Because in most cases the quantum yield ϕ is much less than 1, the numerator of the left side of eqn (6.13) can be approximated to 1. Then, the following relationship

$$\phi = \sqrt{\frac{k_e k_h[A_{ad}][D_{ad}]}{k_r \alpha I}} \tag{6.14}$$

or

$$\frac{1}{\phi} \propto \sqrt{I} \tag{6.15}$$

is derived. Therefore, the plot of the reciprocal of the quantum yield against the square root of the photo-intensity I becomes linear. Figure 6.3 shows that the experimental values of the yield of acetaldehyde formed by the oxidation of methanol fulfill the relationship of eqn (6.15).[5]

According to eqn (6.13), when the photo-intensity is extremely small, namely in the extreme of $I \to 0$, $\phi \to 1$ holds always. However, a problem arises because in order to introduce eqn (6.13), electrons and holes in the particles are prerequisitely treated like the reactive

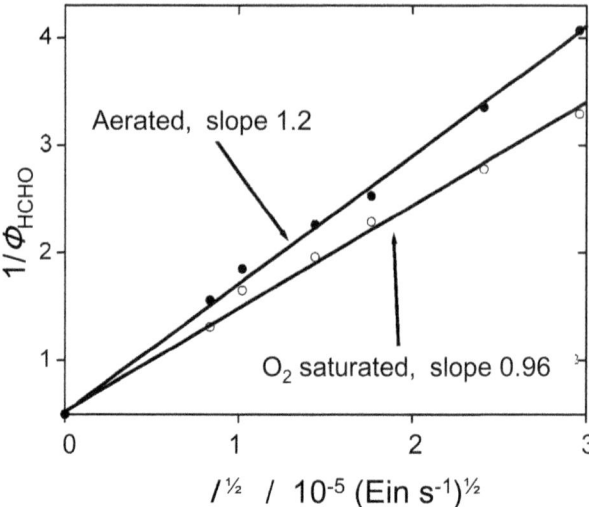

Figure 6.3 The dependence of quantum yield of acetaldehyde from methanol on light intensity. Ein represents the Einstein for the unit of photo-intensity.
Reprinted with permission from Y. Du and J. Rabani, *J. Phys. Chem. B*, 2003, **107**, 11970. Copyright 2003 American Chemical Society.[5]

molecules in the homogeneous solution. The electrons and holes which cause the photocatalytic reactions exist not in the solution but in the limited space in the nanoparticle with the density of N_0. Then, reconsider eqn (6.13) on the basis of the reaction model by taking the particle size into account.

6.2.3 Effect of Particle Size

When the ratio of the particles which possess n electrons and m holes in the volume V is defined X_m^n, the transition between the particles can be depicted as a two-dimensional ladder as shown in Figure 6.4A. Eqn (6.5) to (6.8) can be replaced by the following equation,

$$\frac{dX_m^n}{dt} = (n+1)k_e X_m^{n+1} + (m+1)k_h X_{m+1}^n + (n+1)(m+1)k_r X_{m+1}^{n+1}$$

$$+ g(t)V X_{m-1}^{n-1} - \{nk_e + mk_h + nm(k_r/V) + g(t)V\}X_m^n$$

$$(6.16)$$

The quantum yield of the reduction product A^- corresponding to eqn (6.9) can be calculated with the following equation.

$$\phi = \int \sum_n \sum_m nk_e X_m^n dt \Big/ \int g(t)V dt \qquad (6.17)$$

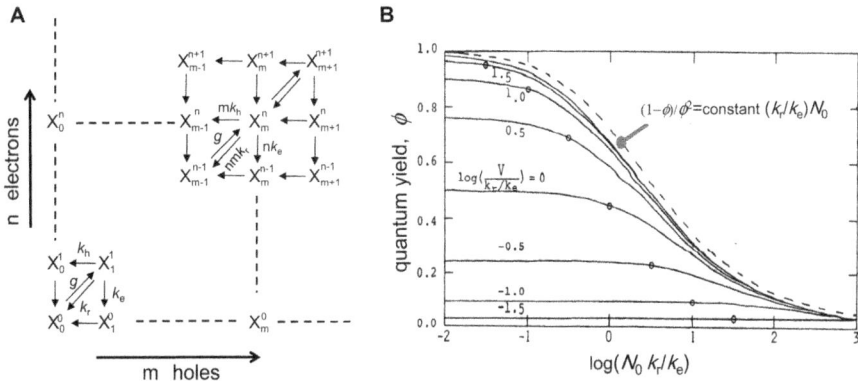

Figure 6.4 (A) Two-dimensional ladder kinetic model for the photoinduced reaction at ultrasmall semiconductor. X_m^n represents the fraction of a particle having n electrons and m holes, which are both induced by photon absorption. Transitions are represented by eqn (6.16). (B) Relationship between the quantum yield for electron transfer Φ and the normalized absorbed photon density $N_0 k_r / k_e$ for nano particles of various volumes (V). Reprinted with permission from Y. Nosaka, N. Ohta and H. Miyama, *J. Phys. Chem.*, 1990, **94**, 3752. Copyright 1990 American Chemical Society.[6]

When at $t = 0$, the initial condition is set to $X_0^0 = 1$ and $X_m^n = 0$, ϕ can be numerically calculated. The results are shown in Figure 6.4B.[6] The broken line in the figure is obtained with eqn (6.13). When V is smaller than k_r / k_e, the size effect of the nanoparticle takes place and ϕ does not become close to 1 but takes a certain value of ϕ_0. Actually, for the research on the light absorption and the oxidation of 2-propanol under the weak light irradiation ($36\ \mathrm{nW\ cm^{-2}}$) on the TiO_2 surface, the limit value of 0.28 was obtained as ϕ.[7] Namely, in this case the condition that the volume of the particles is as small as that comparable to the ratio of rate constants k_r / k_h is fulfilled.

Let's estimate the number of photons in a particle. A TiO_2 anatase particle of 20 nm size ($a = 10$ nm) has a volume V of $4.2 \times 10^{-18}\ cm^3$. When the irradiated UV light has a power of $1\ \mathrm{mW\ cm^{-2}}$ at 360 nm, the photon flux I is calculated to be $1.8 \times 10^{15}\ \mathrm{cm^{-2}\ s^{-1}}$. Because the absorption coefficient α of anatase TiO_2 at 360 nm is about $10^3\ cm^{-1}$, the generation rate of electron-hole pairs, g ($= \alpha I$), is $1.8 \times 10^{18}\ \mathrm{cm^{-3}\ s^{-1}}$. The number of electron-hole pairs formed in the particle per unit time is calculated to be $gV = 7.6\ s^{-1}$. Therefore, if the reaction had not finished within 130 ms ($= 1/gV$), the electron-hole pair generated by the next absorption would affect the yield.

In the general experiments on photocatalysis, the decomposition rate or the decomposed amount are mostly measured but not the quantum yields. Hence from now on the discussion will proceed on the reaction rates instead of on the quantum yields.

6.2.4 Decomposition Rate of Reactant

On definition of the rate of photocatalytic reactions R_D as the decrease of the reactant, the following relations can be derived from eqn (6.9) with $g = \alpha I$.

$$R_D = \frac{d[D^+]}{dt} = -\frac{d[D]}{dt} = \alpha \phi I \tag{6.18}$$

Because most of the electrons and holes simultaneously diminish, one can assume the relation $[e^-] = [h^+]$. Then, eqn (6.12) becomes eqn (6.19).

$$k_e[A_{ad}] = k_h[D_{ad}] \tag{6.19}$$

For oxidation reactions, the reaction rate R_D in eqn (6.18) is expressed as eqn (6.20) by using eqn (6.14) and (6.19).

$$R_D = k_h[D_{ad}]\sqrt{\frac{\alpha I}{k_r}} \tag{6.20}$$

Thus, the reaction rate for the photocatalysis is proportional to the concentration of the adsorbed reactant and the square root of the photo-intensity.

When the photo-intensity I is small, ϕ in eqn (6.18) becomes constant as $\phi = \phi_0$ as described above. Then,

$$R_D = \alpha \phi_0 I \tag{6.21}$$

In this case, eqn (6.21) indicates that the reaction rate is proportional to the photo-intensity, namely reaction rate is determined by the photo-intensity. This is called a light intensity control reaction. In Figure 6.5A the general scheme which depicts the dependency of reaction rates (R_D) on the intensity of the irradiated light (I) is shown.

6.2.5 Assumption of Adsorption Equilibrium

The concentration of the reactants adsorbed on the surface can be calculated from the adsorption equilibrium as follows. There is an adsorptive site S_D on the catalyst surface, and the equilibrium

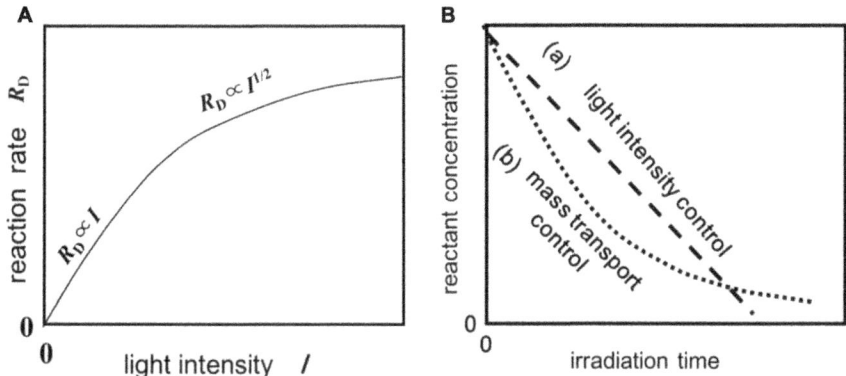

Figure 6.5 Schematic illustration for (A) dependence of initial reaction rate on the light intensity and (B) time dependence of reactant concentration.

between the molecules (D_{ad}) adsorbed on the S_D and those (D) free from the adsorption is expressed as eqn (6.22),

$$D + S_D \xrightarrow{K_D} D_{ad} \tag{6.22}$$

where the equilibrium constant K_D for the adsorption can be expressed as follows.

$$K_D = \frac{[D_{ad}]}{[D][S_D]} \tag{6.23}$$

$[S_D]$ is the concentration of the vacant adsorption site on which D can be adsorbed. When the site concentration before the adsorption of D is expressed by $[S_D]_0$, $[S_D]$ can be expressed as eqn (6.24).

$$[S_D] = [S_D]_0 - [D_{ad}] \tag{6.24}$$

By substituting eqn (6.24) for eqn (6.23), the following relation is obtained.

$$[D_{ad}] = \frac{K_D[S_D]_0[D]}{1 + K_D[D]} \tag{6.25}$$

By using this relation, eqn (6.20), which describes the reaction rate for the oxidation, can be rewritten as follows.

$$R_D = \frac{k_h K_D[D][S_D]_0}{(1 + K_D[D])} \sqrt{\frac{\alpha I}{k_r}} \tag{6.26}$$

On putting the apparent rate constant k_{app} as follows,

$$k_{app} = k_h [S_D]_0 \sqrt{\frac{\alpha I}{k_r}} \qquad (6.27)$$

the reaction rate R_D for the oxidation of D is expressed as follows.

$$R_D = \frac{k_{app} K_D [D]}{(1 + K_D [D])} \qquad (6.28)$$

The reaction mechanism on the assumption of such adsorption equilibrium is called a Langmuir–Hinshelwood mechanism. By taking the reciprocals of both sides of eqn (6.28), eqn (6.29) is obtained.

$$\frac{1}{R_D} = \frac{1}{k_{app}} + \frac{1}{k_{app} K_D [D]} \qquad (6.29)$$

This equation indicates that when the amount of the adsorption is small ($[D_{ad}] \ll [D]$), where $[D]$ could be approximated to the initial concentration $[D]_0$, a linear line is obtained on plotting the reciprocal of the initial reaction rate against the reciprocal of $[D]_0$.

6.2.6 Cases of Light Control and Diffusion Control Reactions

6.2.6.1 Light Intensity Control Reaction

According to eqn (6.28), the reaction rate R_D becomes equal to k_{app} when the adsorption constant K_D is large or the substrate concentration $[D]$ is high, namely $1 \ll K_D[D]$. As a result, the concentration of the substrate decreases lineally against the irradiation time like eqn (6.30) as illustrated in Figure 6.5B(a).

$$[D] = [D]_0 - k_{app} t \qquad (6.30)$$

Such a reaction is called a zero-order reaction because the reaction rate does not depend on the concentration of the reactant. This is the reason why it is called a light intensity control reaction.

6.2.6.2 Mass Transfer Control Reaction

On the other hand, in the case of the high concentration of the substrate $[D]$ and/or small adsorption constant K_D, *i.e.*, when $1 \gg K_D[D]$ holds, eqn (6.28) becomes as follows.

$$R_D = -\frac{d[D]}{dt} = k_{app} K_D [D] \qquad (6.31)$$

By integrating this equation, the time dependence of the reactant concentration [D] is obtained.

$$[D] = [D]_0 \exp(-k_{app}K_D t) \tag{6.32}$$

Therefore, the reactant decreases exponentially according to the first-order reaction process.

For the actual reaction system, the initial stage of the reaction is light intensity control, but when the amount of the reactive materials becomes small, mass transfer control becomes the case. Sometimes one cannot discriminate one from the other. Figure 6.5B(b) shows the change of the concentration of the reactants for the mass transport control reaction (first-order reaction). In many literatures, the first-order process is initially assumed, and the plot of the following equation is used with a first-order rate constant k.

$$\ln \frac{[D]}{[D]_0} = -kt \tag{6.33}$$

6.2.6.3 Example of the Cases

Let's take 2-propanol as an example to see how the concentration of the reactants is actually related to the mass transport control and the light intensity control reactions.[8] The region in Figure 6.6 where the amount of light (the intensity of the UV light) is low and the concentration of the reactants is high (the region surrounded by the curve B at the lower right site) is the region for the light intensity control reactions. In this region, the amount of the light dominates the reaction rates. Because the amounts of the adsorption and the light absorption are proportional at the boundary of this region, curve B is similar to that corresponding to the adsorption isotherm.

On the other hand, the region where the amount of light is high and the reactant concentration is low (the region surrounded by the curve A at the higher left site) is the region for the mass transport control reactions. In this region, because the rate at which the reactant molecules diffuse toward the surface controls the reaction, different from the case for the light intensity control process, it is affected by the flow velocity of the reactants in the reaction chamber. On increasing the flow velocity of the reactant by 100 times, from $1\,cm\,s^{-1}$ to $100\,cm\,s^{-1}$, the lowest light intensity of mass transfer control reactions also becomes about 100 times larger (curve A'). The lower limit of curve A, *i.e.*, the lower limit of the intensity of ultraviolet

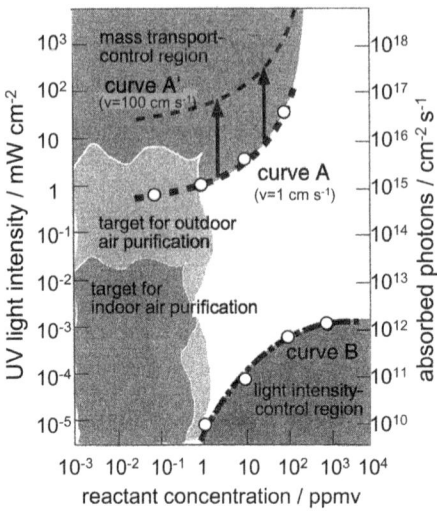

Figure 6.6   Illustration of regions of pure mass transport control conditions and pure light intensity control conditions in the photocatalytic degradation of gas-phase organics on a plot of light intensity *vs.* initial reactant concentration: curves A and A', boundary of the mass transport control; curve B, boundary of the pure light intensity control region.[8]
Reprinted with permission from Y. Ohko, A. Fujishima and K. Hashimoto, *J. Phys. Chem. B*, 1998, **102**, 1724. Copyright 1997 American Chemical Society.

light for mass transport control rate, is 1 mW cm^{-2}, which is lower than the UV light intensity (3 mW cm^{-2}) of the sunlight in the summer time. This indicates that the outdoor air cleaning by TiO_2 proceeds by the mass transport control rate. In other words, it indicates that as a light source for outdoor air cleaning, the sunlight possesses enough light intensity, and that the condensation of the light is not so effective. For instance, from the figure, in the case of the air flow of 1 cm s^{-1} the light intensity effective to decompose the reactant of 0.1 ppmv can be estimated as 0.5 mW cm^{-2}.[8]

6.2.7   Effect of Temperature

The temperature dependency of the reaction rate for photocatalytic reactions is complicated. In eqn (6.27), the apparent oxidation rate constant k_{app} is expressed by two rate constants, k_h and k_r. Generally the rate constant for an elementary reaction is given by,

$$k = k_0 \exp(-E_a/k_B T) \tag{6.34}$$

where E_a is the activation energy ($E_a > 0$) and k_B is Boltzmann's constant. Therefore, the effective rate constant k_{app} ($\propto k_h/\sqrt{k_r}$) increases with the increase of temperature T when E_a for k_h is larger than half of that for k_r. Accordingly, the photocatalytic reactions are considered to become rapid.

Though k_e, k_h, and k_r contain several stages, the behaviors of electrons and holes generated by the light in the semiconductors could be depicted as shown in Figure 6.2. The generation rate of electrons and holes is determined by the rate of photo-absorption. Because the absorbance of a semiconductor at the energy near the bandgap edge is increased with temperature as shown in Figure 2.6A, the generation rate, or αI in eqn (6.27), increases with the temperature. Though it does not take time for photoinduced electrons and holes to reach the surface of nanoparticles, the recombination of electron-hole pairs in the particle which affects k_r may increase with temperature.

The photocatalytic reaction rate R_D is also affected by the association constant K_D as in eqn (6.31) under the assumption that the reaction takes place in the adsorbed state. Usually the equilibrium constant K_D obeys the following eqn (6.35),

$$K_D = K_0 \exp(-\Delta H/k_B T) \qquad (6.35)$$

where ΔH is the enthalpy change with the adsorption. Because in general adsorption is an exothermic reaction, $\Delta H < 0$ holds and the equilibrium constant becomes smaller with the increase of temperature. Thus, the photocatalytic reaction rate R_D is considered to become lower with increasing temperature.

There are many factors affecting the temperature dependency of the photocatalytic reaction rate as stated above. Hence, different from general chemical reactions, for photocatalysis it does not always hold that on increase of temperature the reactions are accelerated.

6.3 Contribution of OH Radicals to the Reaction Mechanism

6.3.1 Possible Pathways of Photocatalytic Oxidation

Although OH radials are generally believed to be the most important active oxygen species, the contribution to the photocatalytic reactions is not always large on investigating the previous researches. Nevertheless, it appears that many literatures easily ascribe the photocatalytic reactions to the contribution of OH radicals. Then, let's consider the

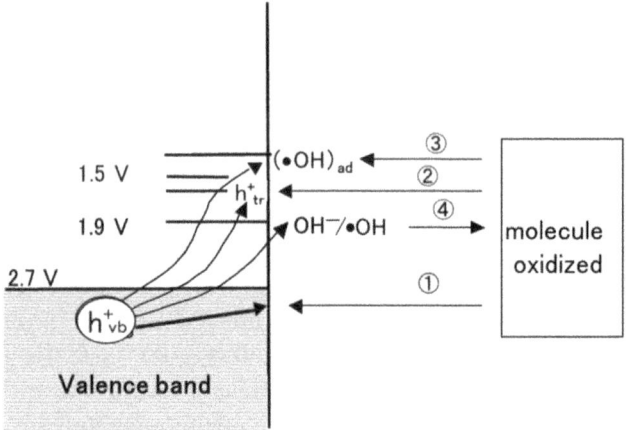

Figure 6.7   Plausible oxidation mechanisms for the oxidation of a molecule by a photogenerated valence band hole. Arrows indicate the transfer of the positive hole and the molecule.

contribution of OH radicals to the oxidation decomposition reactions. The plausible oxidative decomposition procedures in photocatalysis would be following four paths shown in Figure 6.7.

① The valence band holes (h^+_{vb}) directly extract the electron from the adsorbed molecules and oxidize them.

② The valence band holes are stabilized on the TiO_2 surface to become trapped holes (h^+_{tr}), which oxidize the molecules at the surface.

③ The valence band holes oxidize the surface hydroxyl groups or adsorbed waters to generate adsorbed OH radicals ($^\bullet OH_{ad}$), and then they oxidize the molecules at the surface.

④ The generated OH radical is released from the surface to the atmosphere or into the solution. Then, the OH radicals can oxidize the compounds distant from the surface.

The potentials in the figure are those suspected at neutral pH. The redox potential of $^\bullet OH + e^- \rightarrow OH^-$ is 1.9 V (*vs.* SHE) from Figure 4.5A.

The redox potential of the adsorbed OH radical which is generated by irradiating waters with electron pulse and then adsorbed on the TiO_2 is obtained to be 1.5 V.[9] But, the authors stated that OH radicals adsorbed on the TiO_2 could not be discriminated from the trapped holes. On the basis of this statement, it is often regarded that the tapped holes are actually the OH radicals adsorbed on the TiO_2. However, the signal of OH radicals could not be detected even by the low-temperature ESR method as stated previously. Therefore, it may become an $^\bullet O^-$ radical because at higher pH $^\bullet OH$ becomes $^\bullet O^-$ in homogeneous solution. It is not strange that the adsorption on the Lewis acid point of TiO_2 surface takes place in the form of $^\bullet O^-$ in neutral pH solution. Therefore, taking into account the fact that adsorbed OH radicals were not observed by ESR, the process ③ does not occur but process ② would be regarded to take place.

It is controversial whether OH radicals are involved in photocatalytic reactions or not. There is a similar research on the low-temperature measurements of methanol.[10] In this case, the radical ($^\bullet CH_2OH$) formed by abstracting a hydrogen atom from the methyl group was observed but the OH radical and the radical formed by cleaving the Ti–OCH_3 were not observed, indicating that the Ti–O bond is hardly cleaved. This fact also supports the experimental result that the OH radical is not formed from the terminal OH (Ti–OH). For the photocatalytic decomposition at room temperature, the CH_2OH radical could be observed but not $CH_3O^\bullet$.[11] On summarizing the above experimental results, it can be concluded that photocatalytic oxidation proceeds at room temperature not *via* OH radicals. However, OH radicals with a yield of one-hundredth of trapped holes would be generated for anatase TiO_2 as will be described below. Therefore, it is considered that the oxidation *via* OH radical is not the main procedure but actually may take place for non-adsorbed molecules.

6.3.2 Oxidation Mechanisms with OH Radicals

Next, take a glance at the reactions *via* the OH radicals on the basis of the OH radical detection. The oxidation reaction mechanisms of organic additives (A), such as alcohols (methanol and ethanol) and inorganic ions (I^-, Br^-, and SCN^-) in photocatalysis were elucidated by a fluorescence probe method (see Figure 5.13B) by adopting coumarin (Cou) and coumarin-3-carboxylic acid as fluorescence probes.[12] The coumarin-3-carboxylic acid is adsorbed on TiO_2 surface more strongly than alcohols. Therefore, it can trap only the OH radicals near the surface.

As shown in Figure 6.8 for the reaction scheme, the reactivity of OH radicals to some additive A can be examined by measuring the generation rate of OH radical adducts (OH-Cou) by changing the concentration of A. The trapped holes are firstly produced at the formation rate of g' as eqn (6.36), then the OH radical is generated with the rate constant k_0 as eqn (6.37). When the additive reacts with

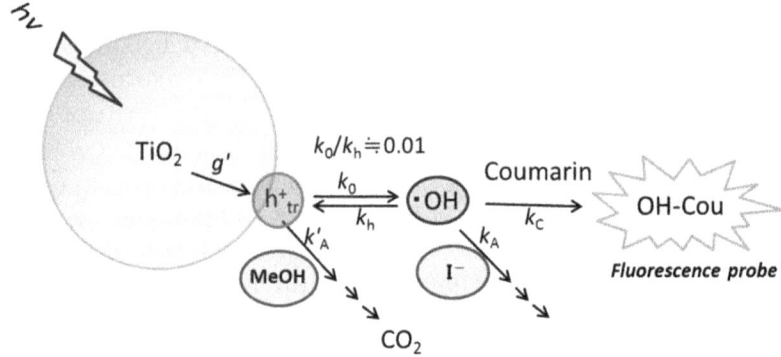

Figure 6.8 Plausible photocatalytic oxidation processes for methanol and iodide ions deduced from the OH radical detection with a coumarin fluorescence probe.
Reprinted from J. Zhang and Y. Nosaka, Photocatalytic oxidation mechanism of methanol and the other reactants in irradiated TiO$_2$ aqueous suspension investigated by OH radical detection, *Appl. Catal. B: Environ.*, **166**, 32–36. Copyright (2015) with permission from Elsevier.

OH radicals or trapped holes as eqn (6.38) or (6.38′), the generation of the fluorescent product of OH-Cou represented by eqn (6.39) is decreased.

$$TiO_2 + hv \xrightarrow{g'} h_{tr}^+ \tag{6.36}$$

$$h_{tr}^+ + H_2O \xrightarrow{k_0} {}^{\bullet}OH + H^+ \tag{6.37}$$

$${}^{\bullet}OH + A \xrightarrow{k_A} product \tag{6.38}$$

$$h_{tr}^+ + A \xrightarrow{k_A'} product \tag{6.38′}$$

$${}^{\bullet}OH + Cou \xrightarrow{k_C} \alpha \cdot OH\text{-}Cou \tag{6.39}$$

The rate constant k_A of eqn (6.38) was obtained from the experiments and then it was significantly smaller than that reported in the literature in homogeneous solution. Thus, it is clear that alcohols react with trapped holes. On the other hand, for halide ions the experimentally obtained k_A was comparable to that in the literature.[12] Thus halide ions react with OH radicals far from the surface, because alcohols can be adsorbed on the TiO$_2$ surface but the halide ions with negative charge do not adsorb on the surface.

For alcohols, from eqn (6.36), (6.37), (6.38′), and (6.39), the decrease of the generation rate of OH-Cou by the addition of A should not depend on the concentration of Cou for the decrease of the OH-Cou generation rate. However, the experimental results indicate that it decreases with increasing the concentration. Hence, eqn (6.40) which describes the transfer of an OH radical to a trapped hole must be assumed. By adopting eqn (6.40), the dependency of the generation rate of OH-Cou on the concentration of A is expressed as eqn (6.41).

$$\bullet OH + TiO_2 \xrightarrow{k_h} h_{tr}^+ + OH^- \tag{6.40}$$

$$\frac{1}{\dfrac{d[OH\text{-}Cou]}{dt}} = \frac{k_0}{g'\alpha} + \frac{k_A'(k_h + k_C[Cou])[A]}{g'\alpha k_C[Cou]} \tag{6.41}$$

Under the assumption of the reaction constant of diffusion limit, $k_A' = 1 \times 10^{10}$ M^{-1} s^{-1}, k_0/k_h becomes around 0.01, indicating that 1% of the trapped holes are in rapid equilibrium with OH radicals in solution.[12]

6.4 Decomposition of Organic Compounds

Primary reaction pathways for the oxidation are usually the direct reactions at the surface of TiO$_2$ with valence band holes or trapped holes, which correspond to path ① or ② in Figure 6.7. Because it is generally known that the photocatalytic oxidation of organic compounds is accelerated by oxygen,[13] the produced radical may react with the reduction products of O$_2$, namely $\bullet O_2^-$ and H$_2$O$_2$. But, the molecular oxygens in air can directly react with the radicals produced by the photocatalytic oxidation because a kind of chain reaction with O$_2$ starting from organic free radicals is well known as auto oxidation.[14] The consumption of O$_2$ at the oxidation site of the photocatalyst has been suggested from the experiment of electrochemical probe reactions at the surface of an illuminated TiO$_2$ photoelectrode.[15] Therefore, the generalized reaction mechanism of the photocatalytic oxidation of organic molecules (RH) can be illustrated as Figure 6.9. RH will degrade by losing one carbon atom by releasing CO$_2$ to generate R′H, but the intermediates may be aldehyde R′CHO or carboxylate R′COO$^-$. The primary reaction mechanisms for some representative organic substances are as follows.

Methane: The most simple organic compound, methane, which can selectively convert into methanol over TiO$_2$ (rutile) under the

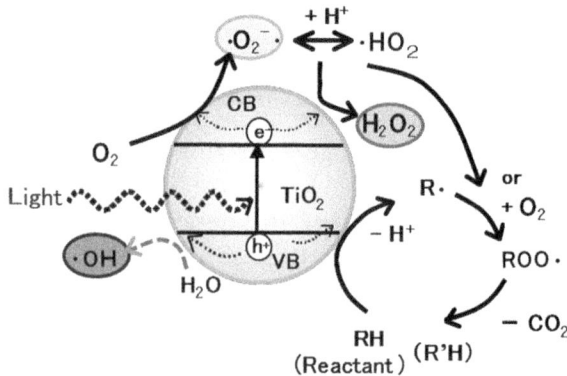

Figure 6.9   General scheme of TiO$_2$ photocatalysis for the decomposition of organic molecules.

irradiation of a strong laser light in aqueous suspension.[16] The formation amount of by-product H$_2$ is small.

Methanol: In a gas flow reaction system, bare-TiO$_2$ does not decompose methanol due to the fast production of formate from the generated methoxide that poisons the active centers.[17] ESR technique[10] revealed the formation of $\cdot$CH$_2$OH resulted from the oxidation of chemisorbed methanol on the TiO$_2$ surface. Figure 6.10 shows that the radical rapidly becomes the adsorbed formaldehyde, and then is decomposed to CO$_2$ and H$_2$O, which was revealed by infrared (IR) studies.[18]

Ethanol: A study by *in situ* IR spectroscopy indicated that the coverage of ethanol and water was important for the initiation of the reaction. The low ethanol coverage on the H$_2$O$_{ad}$-containing TiO$_2$ surface produced adsorbed formate (HCOO$^-_{ad}$) as a primary intermediate, while the high ethanol coverage produced adsorbed acetate (CH$_3$COO$^-_{ad}$) as a major intermediate, resulting in blocking the access of O$_2$ to suppress the reaction.[19]

2-Propanol: Photooxidation of 2-propanol selectively produces firstly acetone. The conversion of 2-propanol to acetone requires two oxidative steps. The first step is aldol condensation on the surface, which has been identified by means of solid-state NMR technique.[20] Only strongly adsorbed reactant species are involved in the photocatalytic oxidation of 2-propanol.[21] In the presence of O$_2$, 2-propanol converts thermally to acetone.[22]

Acetone: The initial step in the acetone oxidation was studied by using ESR at low temperatures.[23] At 100 K, the hole transfers to the adsorbed acetone followed by O$_2$ addition to form an intermediate radical CH$_3$COCH$_2$OO$^\bullet$.

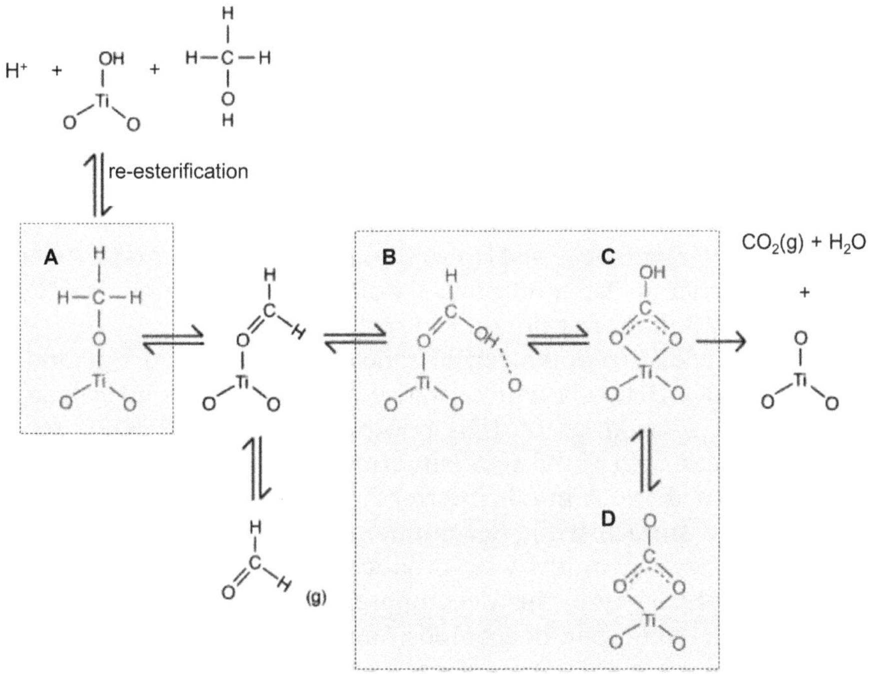

H⁺ + OH (–Ti, O, O) + H–C–H ...

re-esterification

A H–C–H ... ⇌ ... ⇌ B ... ⇌ C ... → CO₂(g) + H₂O +

(g)

D

Figure 6.10 Proposed mechanism for conversion of methanol to CO_2 and water. Gray boxes indicate the surface species characterized by FTIR.
Reprinted with permission from J. R. S. Brownson, M. I. T. Tejedor and M. A. Anderson, *J. Phys. Chem. B*, 2006, **110**, 12494. Copyright 2006 American Chemical Society.[18]

Acetaldehyde: In the photocatalytic oxidation of acetaldehyde, ESR measurements at 100 K revealed the presence of peroxyacyl species ($RCO_3^•$), which are known as oxidative intermediates in the auto-catalytic oxidation of aldehydes.[24]

Formic acid: The photooxidation of formate on TiO_2 surface can be seen in the oxidation process of methanol.[18] In aqueous suspension, carboxyl anion radicals $^•CO_2^-$ were detected by a DMPO spin-trapping ESR method as the intermediate in the decomposition of formate.[25]

Acetic acid: The flow ESR measurements allowed the direct detection of intermediate radicals $^•CH_3$ and $^•CH_2COOH$ in aqueous solution in the absence of O_2.[26] A dehydrated surface or surface that contains fewer terminal OH groups produces $^•CH_3$ more preferably, indicating direct oxidation of the adsorbed acetic acid. In the presence of O_2, these radicals could not be detected at room temperature, suggesting the rapid reaction with O_2.

Methylamines: The H-atom abstraction from $(CH_3)_4N^+$ by $^\bullet OH$ initiated successive demethylation processes to generate tri-, di-, and monomethyl ammonium/amine as an intermediate and NH_3/NH_4^+ as a final product. The photocatalytic degradation rates of $(CH_3)_4N^+$ were comparable in both acidic and alkaline conditions, which could not be explained by a simple electrostatic surface charge model. By using $^\bullet OH$-scavenger, *tert*-butyl alcohol, as a diagnostic probe of the mechanism, it is suggested that the photocatalytic oxidation of $(CH_3)_4N^+$ under acidic conditions should proceed through free $^\bullet OH$ in the bulk solution, but not on the surface of TiO_2.[27]

Amino acids: Photocatalytic decomposition of alanine, $H_2NCH(CH_3)COOH$, as a representative amino acid was investigated in detail for nine kinds of TiO_2 powders by using ^1H NMR spectroscopy. Acetic acid is the stable intermediate *via* acetaldehyde as a fragile intermediate. A minor intermediate pyruvic acid was formed on the rutile surface while acetamide was formed on the anatase surface.[28] The acetamide was produced from acetaldehyde and ammonia, which were the decomposed components of alanine. Furthermore, the photodecomposition rates of seven amino acids on the anatase TiO_2 surface were elucidated by ^1H NMR.[29] The decomposition rates increased in the order of Phe < Ala < Asp < Trp < Asn < His < Ser, which were correlated with the changes in the iso-electric point on adsorption. Because the iso-electric point shifted to a lower pH with increasing decomposition rates for Phe, Trp, Asn, His, and Ser, the effective adsorption and photocatalytic sites for these amino acids are considered to be the basic terminal Ti-OH groups.[29]

Peptides: Decompositions with TiO_2 powder of glutathione and related amino acids, Glu, Cys, and Gly, were measured by ^1H NMR spectroscopy.[30] The results suggest that both glutathiones in reduced and oxidative forms should be adsorbed on the TiO_2 surface by carboxyl or amino groups but not by the thiol group of the side chain which plays a crucial role in the glutathione cycle, to be degraded.[30]

Benzene: For the oxidation of benzene, the marked difference in photocatalytic activity between rutile and anatase powders was observed, which was ascribed to the involvement of O_2.[31] On anatase particles, benzene is oxidized to phenol efficiently through an oxygen-transfer process using water as the oxygen source, as shown in Figure 6.11A. The atomic configuration on the surface of the anatase particles is favorable for their efficient production using water as the oxygen source. Anatase particles have more irregular structures on the surface, which may contribute to the formation of $Ti-O^\bullet$, or $Ti-OO^\bullet$

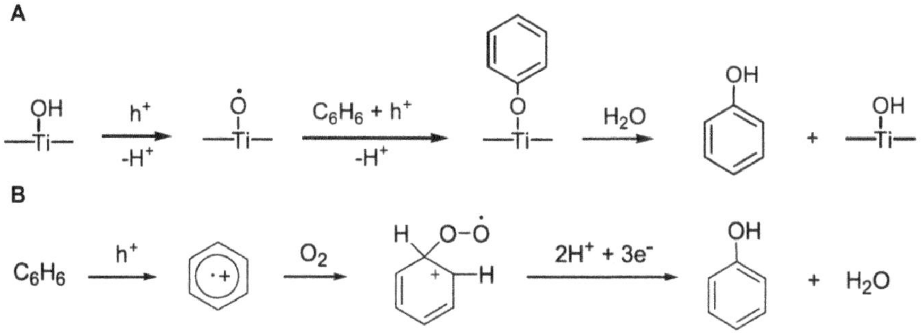

A

B

Figure 6.11 Plausible mechanism for production of phenol from benzene through (A) oxygen transfer process using water as oxygen source and (B) hole transfer process using O_2 as oxygen source.
Reprinted with permission from T. D. Bui, M. Matsumura, *et al.*, *J. Am. Chem. Soc.*, 2010, **132**, 8453. Copyright 2010 American Chemical Society.[31]

that is used to accelerate the oxygen-transfer process from water. On the other hand, in the case of rutile particles as shown in Figure 6.11B, the contribution of the oxygen-transfer process is small and the hole transfer process becomes dominant. The availability of the oxygen-transfer process reflects the higher photocatalytic activity for anatase particles.[31]

Toluene: Toluene reacted quickly to form benzaldehyde, which was also oxidized rapidly to form less-reactive intermediates. These intermediates were accumulated on the surface and reduced the photocatalytic oxidation rate. When the feed was humidified, the deactivation became slower.[32] However, the high humidity in the feed stream significantly reduced the oxidation rate of toluene because water was preferentially adsorbed on the hydrophilic surface of TiO_2.[33]

Phenols: In phenol photooxidation, the major intermediates are 1,4-dihydroxybenzene (hydroquinone) and 1,2-dihydroxybenzene (catechol). The product analysis for six monosubstituted benzenes (substituted with $-OH$, $-NH_2$, $-NHCOCH_3$, $-NO_2$, $-CN$, and $-COCH_3$) proved that the reaction of monohydroxylation of an aromatic ring occurred by •OH attack in the reaction.[34] However, because the mechanism of the hole oxidation has not been clearly understood, the hole oxidation mechanism cannot be discarded.

Pyridine: Analysis of the photocatalytic process for pyridine degradation is shown in Figure 6.12. The major degradation products for pyridine are acetate ion, formate ion, and ammonium ion. But on

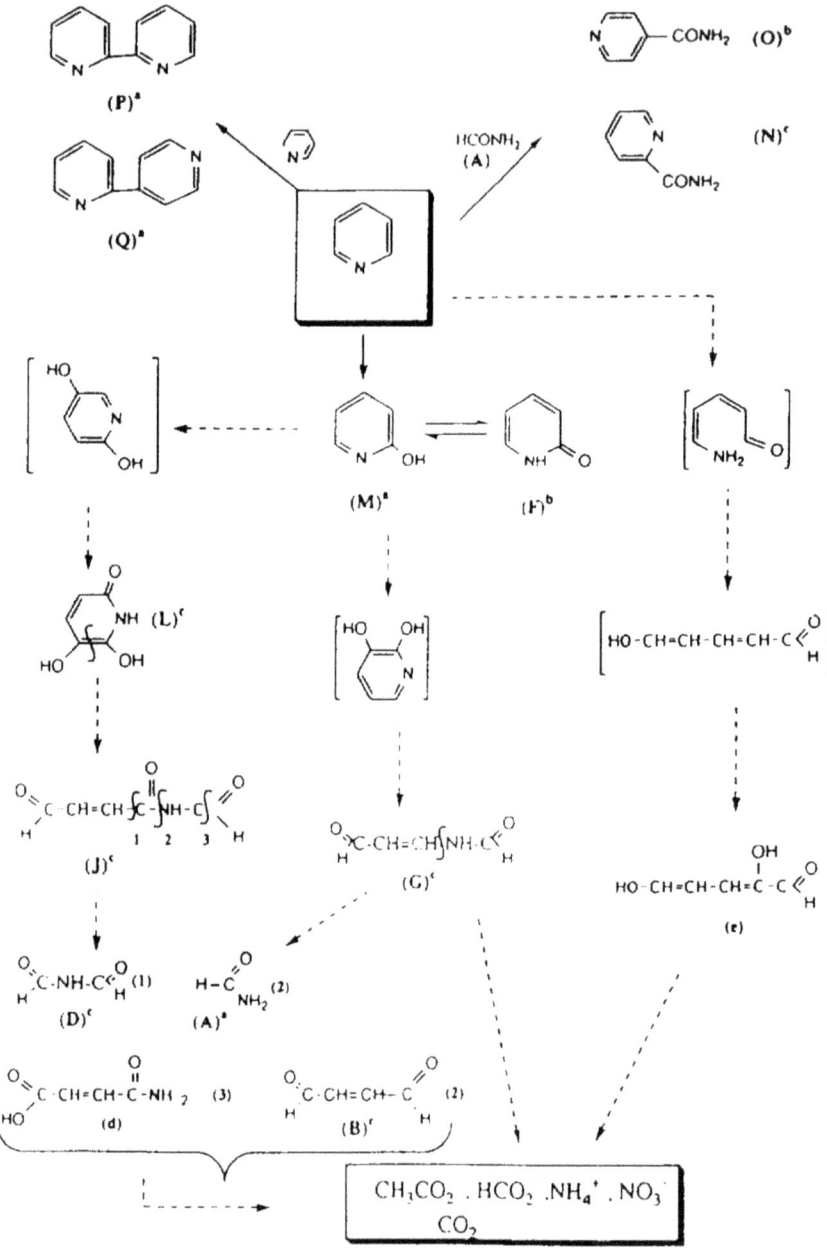

Figure 6.12 Decomposition process of pyridine with a TiO$_2$ photocatalyst. Reproduced from P. Pichat, Photocatalytic degradation of aromatic and alicyclic pollutants in water: by-products, pathways and mechanisms, *Water Sci. Technol.*, 1997, **35**(4), 73–79, with permission from the copyright holders, IWA Publishing.

analysis of the intermediates, the form of –CH=CH– can be recognized in some intermediates, and more than one C=O groups are attached with all the intermediates.[35] Nitrogen exists as a form of an amino group, but not an imino group. Furthermore, the fact that methyl groups of acetate ion and ammonium ion are formed indicates that the addition of a hydrogen atom takes place during the reaction processes. The formation of dipyridyl can be explained by the formation of pyridinyl radicals formed by dehydrogenation.

Alkyl halides: In the case of alkyl halides, the reduction in place of the oxidation is a key step in the photocatalytic degradation. For the decomposition of CCl_4, dichlorocarbene CCl_2 was formed by two-electron reduction.[36] Because dichlorocarbene readily undergoes base-catalyzed hydrolysis, pH is a critical experimental variable that controls the degradation. For the decomposition of dichloromethane (CH_2Cl_2) in the gas-phase, in addition to the major species HCl and CO_2 significant quantities of phosgene $(COCl_2)$, CO, Cl_2, and CCl_4 were produced. The rate of CH_2Cl_2 degradation presented pronounced oxygen dependence, implying significantly rapid degradation at relatively high O_2 concentrations.

6.5 Antibacterial Exertion

The extinction of bacterial cells by the exertion of TiO_2 photocatalysts has already been reported in the mid 1980s. So far it is recognized that the reaction is initiated with the destruction of the cell membrane, leading to the extinction of cells.[37] By photocatalysis endotoxin, a kind of lipopolysaccharide which forms the cell walls of the Gram-negative bacteria,[38] was decomposed and lipid membranes were peroxidized.[39]

In the case of *Escherichia coli* (*E. coli*), the cell walls are damaged at first and the damage of the cell membranes succeeds, then direct attack by the photocatalysts becomes possible. The smaller the particle size of TiO_2, the faster the damage takes place in the cells.[40] By observing *E. coli* on the TiO_2 thin films by AFM, the substances in the cells such as potassium ions were found to elute as a result of the destruction of the cell walls and the cell membrane, leading to the death of the cells.[41] Furthermore, it was observed that the cell division was caused at several parts of the cells as a result of the damage of the cell walls by the electron microscope.[42] The extinction of *E. coli* by photocatalysts proceeds by two steps, *i.e.*, rapid and slow reactions. For the bacteria without cell walls only the rapid reactions were recognized. The outer membrane of the cell walls plays a role as a barrier but peptidoglycan does not. The outer membrane is destroyed partly to cause the successive disorder of the cell membranes, leading to the death of the cells.[43] Figure 6.13 shows schematically how TiO_2 solely or with the aid of Cu^{2+}, which has sterilizing power, attacks the cells. H_2O_2 is believed to be the active species involved in the attacks. The extinction of phage and virus is considered to proceed also by similar mechanisms.[44]

6.6   Water Splitting

The generation of hydrogen and oxygen by reduction and oxidation of water, respectively by the use of photocatalysts, is an important research subject because by using solar energy water can be directly converted to energy for hydrogen fuel without causing any environmental pollution. The deposition of appropriate catalysts of fine particles such as platinum on the TiO_2 surface makes the reduction of water feasible because the energy of the conduction band edge of TiO_2 is not different largely from that of the reductive potential of water. Details about the platinum deposition will be described in Chapter 7. In this section, the oxidation reactions of water on the TiO_2 surface will be described.

In the photocatalytic reactions of TiO_2, the OH radical has been considered to be generated on the oxidation of water. Hence many researches have taken account of the fact that O_2 is generated *via* H_2O_2, which is produced by the dimerization of OH radicals. As stated previously, in spite of the enormously large number of reports on the oxidation by photocatalysts, there is not a report that demonstrates the generation of H_2O_2 from OH radicals. Therefore, it would be reasonable to regard that OH radials are not generated in the oxidation of water. By comparing the PL (photoluminescence) observed at the wavelength of 840 nm with the oxidative current by the electrode reaction at the (100) facet of rutile TiO_2, it was indicated that at this facet the vacant level of the trapped hole involved in the oxygen generation possessed energy of 1.5 eV less than the lowest level of conduction band. Furthermore, with *in situ* observation by IR spectroscopy the proposed mechanism is that the valence band holes are trapped at the bridged oxygen of [Ti-O-Ti], on which simultaneously solvent water dissociatively adsorbs to form [Ti-O•HO-Ti]. Then, two of these radicals combine with each other to form the oxygen molecule.[45] Based on this mechanism, more details of the oxidation

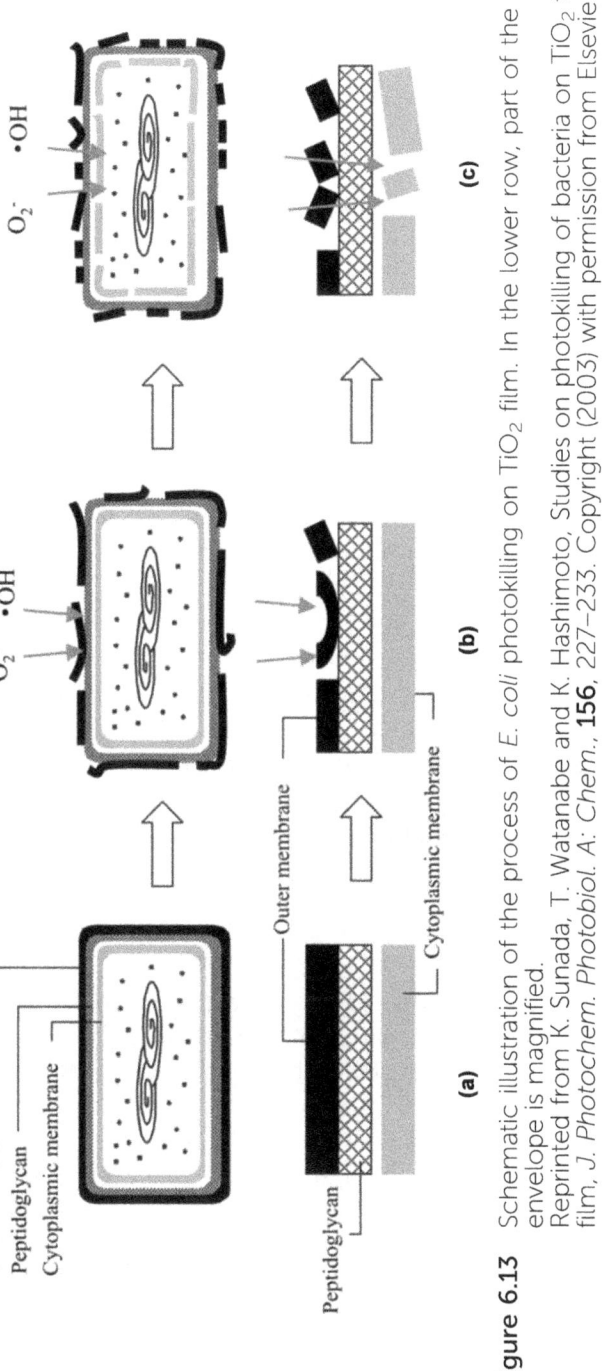

Figure 6.13 Schematic illustration of the process of *E. coli* photokilling on TiO₂ film. In the lower row, part of the cell envelope is magnified. Reprinted from K. Sunada, T. Watanabe and K. Hashimoto. Studies on photokilling of bacteria on TiO₂ thin film, *J. Photochem. Photobiol. A: Chem.*, **156**, 227–233. Copyright (2003) with permission from Elsevier.

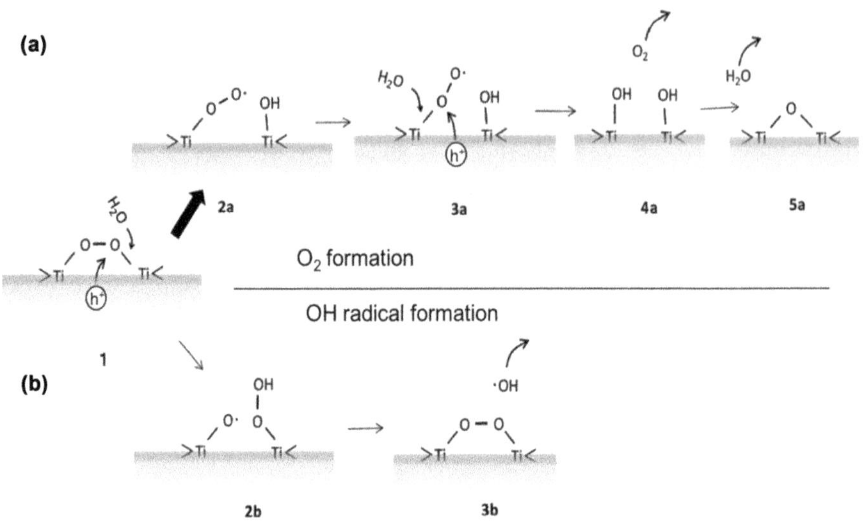

(a)

(b)

Figure 6.14 Plausible reaction steps starting from Ti-peroxo to form (a) O_2 and (b) OH radical at the TiO_2 surface.
Reprinted with permission from Y. Nakabayashi and Y. Nosaka, *J. Phys. Chem. C*, 2013, **117**, 23832. Copyright 2013 American Chemical Society.[46]

mechanism were investigated for TiO_2 single crystal electrodes with (100), (110), and (001) facets. As shown in Figure 6.14, at the third oxidation step among the four oxidation steps, the bond cleaved in the Ti–O–O–Ti group is Ti–O or O–O. This difference results in the formation of O_2 or OH radicals.[46]

6.7 Mechanism of Superhydrophilicity

The surface of glass becomes foggy on exposure to water vapor unless it is coated with TiO_2. Because of the adsorption of water drops of micrometer size on the surface, the light is scattered. Then, the surface of the glass appears foggy. On irradiating ultraviolet light, the TiO_2-coated surface becomes hydrophilic and the water drops spread on the surface to restrain the light scattering. This light induced effect characteristically diminishes several hours after placing the glass in the dark. Phenomena such as the anti-fogging effect of car mirrors and washing out oil stains by the rain are based not only on the properties of decomposition of organic molecules but also on the photoinduced superhydrophilicity.[47] The surface of TiO_2 is not significantly hydrophilic. However, on the illumination of ultraviolet light, the oxygen defects are possibly created, owing to which the surface becomes hydrophilic. This is considered to enhance the dissociative adsorption of water.

However, there is a report that stated that the proposed mechanism to enhance the hydrophilicity is doubtful[48] because it has been recently reported that the hydrophilicity is not caused only by the desorption of the lattice oxygen by the analysis with such TPD and STM methods under ultra-high vacuum. If it is the case, it might be concluded that the surface of TiO$_2$ is intrinsically hydrophilic and that when the impurities on the surface are decomposed by the photocatalytic reactions, the hydrophilicity recovers. However, SrTiO$_3$, which possesses photocatalytic activity, does not present superhydrophilicity. For TiO$_2$, superhydrophilicity decreases by washing the surface with NaOH solution or by ultrasonic treatment of the surface, indicating the presence of a phenomenon other than surface cleaning.[49]

Thus, the superhydrophilicity of an irradiated TiO$_2$ surface is induced as shown in Figure 6.15A by reactions other than photocatalytic decomposition. Actually, the increase in the amount of surface

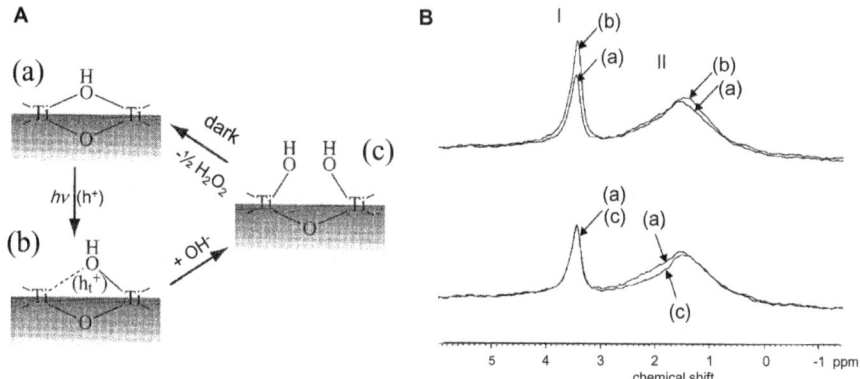

Figure 6.15 (A) (a) (Before UV irradiation) the OH group is bound to an oxygen vacancy, (b) (at the transition state) the photogenerated hole is trapped at the lattice oxygen, and (c) (after UV irradiation) new OH groups are formed. Reprinted with permission from N. Sakai, K. Hashimoto *et al.*, *J. Phys. Chem. B,* 2003, **107**, 1028. Copyright 2003 American Chemical Society.[49] (B) ^{1}H NMR spectra of TiO$_2$ film plate mesured at 800 MHz at 21 °C. (a) Before and (b) just 30 min. after UV irradiation in air; and (c) 2 h stored in the dark after 30 min. UV irradiation in air.
Reprinted with permission from A. Y. Nosaka and Y. Nosaka, *et al.*, *J. Phys. Chem. B*, 2003, **107**, 12042. Copyright 2003 American Chemical Society.[50]

hydroxyl groups by ultraviolet light illumination was observed by IR and NMR spectroscopy. Figure 6.15B shows that two kinds of surface OH groups increase with photo-irradiation and return back almost to the original state after storage in the dark.[50]

It is well known that the wettability of the rough surface is lower than that for the corresponding smooth surface. Although many commercial self-cleaning products have been developed and numerous studies have been performed, the mechanism and many factors influencing the efficiency of superhydrophilicity are still not completely clear. More fundamental understanding is necessary to improve the efficiency and extend the application of this technology to other fields.[51]

6.8 Mechanism of Dye-sensitized Photocatalysis

Dye-sensitized photocatalysis is the reaction utilizing superoxide and hydrogen peroxide generated by the reduction of oxygen on TiO_2 by the photo-excitation of adsorbed dye as shown in Figure 6.16. This reaction is based on the decomposition of the dye D which absorbs light.[52] The characteristic of this technique is that the reaction can be applied to any organic compounds by utilizing visible light when the dye which should be decomposed can absorb the light.

The superoxide radical in this mechanism was directly observed at room temperature for the suspension of TiO_2 sensitized by a porphyrin dye.[53] Furthermore, the TiO_2 film containing copper phthalocyanine prohibits the propagation of *E. coli* by visible light illumination. In this case H_2O_2 and OH radicals are considered to be involved in the reactions to prohibit the propagation.[54]

For dye-sensitized photocatalysis, in most cases the dyes which absorb light are decomposed. However, not much attention is often

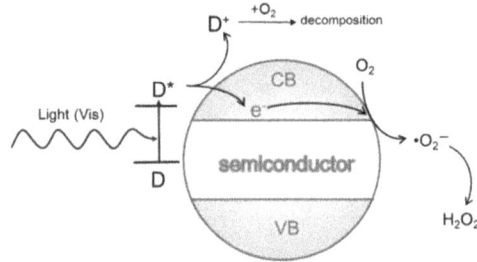

Figure 6.16 Schematic drawing of sensitizer type photocatalytic reaction process.

paid to the fact that this kind of photocatalytic reaction also unconsciously takes place. For instance, when photocatalytic activity is evaluated by measuring the decomposition of the dye, the photocatalytic activity is evaluated from the decrease of the dye, with checking that no change is observed in the absence of TiO_2. However, in this case one cannot deny the possibility that dye-sensitized photocatalytic reactions take place.

If the oxidized dye D^+ in the reaction scheme possesses an oxidative ability without decomposition, the scheme is useful as the photocatalytic system. For example, platinum chloride complex is suggested as the dye to have oxidative ability.[55,56]

References

1. T. Yoshihara, R. Katoh, *et al.*, *J. Phys. Chem. B*, 2004, **108**, 3817.
2. D. W. Bahnemann *et al.*, *J. Phys. Chem. B*, 1997, **101**, 4265.
3. R. Katoh, M. Murai and A. Furube, *Chem. Phys. Lett.*, 2010, **500**, 309.
4. J. Schneider, D. W. Bahnemann *et al.*, *Chem. Rev.*, 2014, **114**, 9919.
5. Y. Du and J. Rabani, *J. Phys. Chem. B*, 2003, **107**, 11970.
6. Y. Nosaka, N. Ohta and H. Miyama, *J. Phys. Chem.*, 1990, **94**, 3752.
7. Y. Ohko, K. Hashimoto and A. Fujishima, *J. Phys. Chem. A*, 1997, **101**, 8057.
8. Y. Ohko, A. Fujishima and K. Hashimoto, *J. Phys. Chem. B*, 1998, **102**, 1724.
9. D. Lawless, N. Serpone and D. Meisel, *J. Phys. Chem.*, 1991, **95**, 5166.
10. O. I. Micic, M.C. Thurnauer *et al.*, *J. Phys. Chem.*, 1993, **97**, 13284.
11. M. Kaise *et al.*, *Langmuir*, 1994, **10**, 1345.
12. J. Zhang and Y. Nosaka, *Appl. Catal., B*, 2015, **166**, 32.
13. A. Maldotti, A. Molinari and R. Amadelli, *Chem. Rev.*, 2002, **102**, 3811.
14. N. A. Clinton, *J. Am. Chem. Soc.*, 1975, **97**, 3757.
15. K. Ikeda, A. Fujishima *et al.*, *J. Phys. Chem. B*, 1997, **101**, 2617.
16. M. A. Gondal *et al.*, *Chem. Phys. Lett.*, 2004, **392**, 372.
17. J. Araña *et al.*, *Appl. Catal. B*, 2004, **53**, 221.
18. J. R. S. Brownson, M. I. T. Tejedor and M. A. Anderson, *J. Phys. Chem. B*, 2006, **110**, 12494.

19. Z. Yu and S. S. C. Chuang, *J. Catal.*, 2007, **246**, 118.
20. W. Xu and D. Raftery, *J. Phys. Chem. B*, 2001, **105**, 4343.
21. F. Arsac, D. Bianchi *et al.*, *J. Phys. Chem. A*, 2006, **110**, 4213.
22. D. Brinkley and T. Engel, *J. Phys. Chem. B*, 2000, **104**, 9836.
23. A. L. Attwood, D. M. Murphy *et al.*, *J. Phys. Chem. A*, 2003, **107**, 1779.
24. C. A. Jenkins and D. M. Murphy, *J. Phys. Chem. B*, 1999, **103**, 1019.
25. L. L. Perissinotti, M. A. Grela *et al.*, *Langmuir*, 2001, **17**, 8422.
26. Y. Nosaka, M. Kishimoto and J. Nishino, *J. Phys. Chem. B*, 1998, **102**, 10279.
27. S. Kim and W. Choi, *Environ. Sci. Technol.*, 2002, **36**, 2019.
28. M. Matsushita, T. H. Tran and Y. Nosaka, *Catal. Today*, 2007, **120**, 240.
29. T. H. Tran, A. Nosaka and Y. Nosaka, *J. Phys. Chem. B*, 2006, **110**, 25525.
30. A. Nosaka, G. Tanaka and Y. Nosaka, *J. Phys. Chem. B*, 2012, **116**, 11098.
31. T. D. Bui, M. Matsumura *et al.*, *J. Am. Chem. Soc.*, 2010, **132**, 8453.
32. M. C. Blount and J. L. Falconer, *Appl. Catal. B*, 2002, **39**, 39.
33. L. Cao, Z. Gao *et al.*, *J. Catal.*, 2000, **196**, 253.
34. G. Palmisano, *Chem. Commun.*, 2006, 1012.
35. P. Pichat, *Wat. Sci. Technol.*, 1997, **35**(4), 73.
36. W. Choi and M. R. Hoffmann, *J. Phys. Chem.*, 1996, **100**, 2161.
37. T. Matsunaga *et al.*, *FEMS Microbiol. Lett.*, 1985, **29**, 211.
38. K. Sunada, K. Hashimoto *et al.*, *Environ. Sci. Technol.*, 1998, **32**, 726.
39. P. C. Maness *et al.*, *Appl. Environ. Microbiol.*, 1999, **65**, 4094.
40. Z. Huang *et al.*, *J. Photochem. Photobiol. A*, 2000, **130**, 163.
41. Z.-X. Lu, D.-W. Pang *et al.*, *Langmuir*, 2003, **19**, 8765.
42. P. Amezaga-Madrid, F. J. Solis *et al.*, *J. Photochem. Photobiol. B*, 2003, **70**, 45.
43. K. Sunada, T. Watanabe and K. Hashimoto, *J. Photochem. Photobiol. A*, 2003, **156**, 227.
44. H. Ishiguro *et al.*, *Appl. Catal. B*, 2013, **129**, 56.
45. R. Nakamura and Y. Nakato, *J. Am. Chem. Soc.*, 2004, **126**, 1290.
46. Y. Nakabayashi and Y. Nosaka, *J. Phys. Chem. C*, 2013, **117**, 23832.
47. R. Wang, A. Fujishima *et al.*, *Nature*, 1997, **388**, 431.
48. J. M. White, J. Szanyi and M. A. Henderson, *J. Phys. Chem. B*, 2003, **107**, 9029.
49. N. Sakai, K. Hashimoto *et al.*, *J. Phys. Chem. B*, 2003, **107**, 1028.
50. A. Nosaka, Y. Nosaka *et al.*, *J. Phys. Chem. B*, 2003, **107**, 12042.

51. L. Zhang, D. Bahnemann *et al.*, *Energy Environ. Sci.*, 2012, **5**, 7491.
52. T. Wu, J. Zhao *et al.*, *J. Phys. Chem. B*, 1999, **103**, 4862.
53. J. Yu, X. Wang, B. Zhang *et al.*, *J. Phys. Chem. B*, 2004, **108**, 2781.
54. J. C. Yu *et al.*, *J. Photochem. Photobiol. A*, 2003, **156**, 235.
55. G. Burgeth and H. Kisch, *Coord. Chem. Rev.*, 2002, **230**, 41.
56. Y. Ishibai *et al.*, *J. Photochem. Photobiol. A*, 2007, **188**, 106.

7 Methods for Improving Photocatalytic Activity

7.1 Downsizing of Particulate Photocatalysts

7.1.1 The Effects of Decrease of Particle size

On decreasing the particle size of semiconductor photocatalysts, as shown in Figure 7.1 the following effects can be caused: ① redox reactions on the surface are accelerated, ② the distance between the photoabsorption position and the surface becomes shorter, and ③ the recombination of photoinduced electron–hole pairs is accelerated.

First of all, it is not limited to the case of photocatalysts, but generally by making the particulate smaller the surface area increases and the number of adsorption sites increases. As stated in Chapter 6, when the number of adsorption sites for oxidation (S_D) and reduction (S_A) increases, the reaction rates on the surface become higher (Figure 7.1①).

Next, as for the quantum size effect stated in Section 3.4, on decrease of the size of particle the energy level of the valence band holes shifts to lower. Therefore, the molecules which cannot be oxidized by usual semiconductor particles could be oxidized by quantized-particles. By this size-quantization, the energy level of the conduction band electrons shifts to the upper accompanied with the shift of the photoabsorption spectrum to the shorter wavelength. Hence, for semiconductors of narrow bandgaps, the bandgap can be controlled by size-quantization in the visible wavelength region.

Introduction to Photocatalysis: From Basic Science to Applications
By Yoshio Nosaka and Atsuko Nosaka
© Yoshio Nosaka and Atsuko Nosaka, 2016
Published by the Royal Society of Chemistry, www.rsc.org

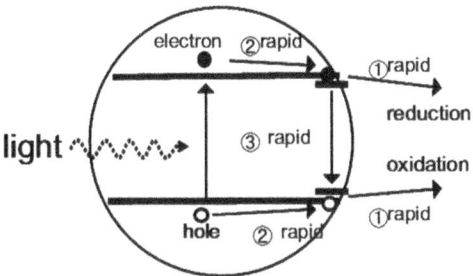

Figure 7.1 Expected effects on semiconductor photocatalysis on the decrease of particle size.

The oxidation reactions are known to become faster by downsizing for TiO_2. However, the energy shift by quantum effect is not large as discussed in Section 3.4, suggesting that the other effect should be responsible for the enhanced reaction. On decreasing the particle size, the conduction band electrons and the valence band holes generated in TiO_2 by photoabsorption begin to locate in the vicinity of the surface. Therefore, the smaller the size of the particles, the faster the carriers reach the surface (Figure 7.1②). When the size becomes as small as about 10 nm, which means that the size is close to that of the exciton, the electrons and holes generated by photoexcitation span to the particle surface. It is indicated that the surface electron density calculated from the expansion of the electron confined in a small size is correlated with the surface electron transfer rate.[1]

It is considered that, similar to the case of semiconductor electrodes, the space charge layer is formed inside the semiconductor photocatalysts and owing to the electric field gradient, the recombination of electrons and holes excited by photons can be avoided. However, for TiO_2, as calculated in Section 4.3.2, the actual thickness of the space charge layer is usually about 100 nm, which is larger than the radius of the usual nanoparticles. Hence, the possibility of the generation of the electric field gradient is low for TiO_2.

Generally the recombination reaction (Figure 7.1③) becomes faster inversely proportionally to the volume or the surface area of the particles as discussed in Section 6.2.3. Namely, the recombination increases because electron–hole pairs are generated in the narrow space. However, because the redox reactions on the surface compete with the recombination reaction, depending on the surface reaction rate and the recombination reaction rate, the reaction efficiency can conversely become lower by making the particle size smaller. Actually, as shown in Figure 6.4, the reaction yield can become lower with a decrease of the particle size (particle volume) under the assumption that the charge transfer reaction rate does not depend on the particle size.[2]

For instance, for rutile TiO_2 photocatalyst of particle size of 12–150 nm, with deposited Pt nanoparticles, hydrogen generation from water and alcohol became larger with decreasing the particle size (see Figure 7.2A).[3] To the contrary, as shown in Figure 7.2B, for TiO_2 fabricated from $TiOSO_4$, on increasing particle size up to 6–32 nm, the reaction rate was increased. This was mainly explained not only by the decrease of the recombination rate due to the increase of the volume but also the increase of the surface defects by calcination.[4]

The coordination states of the titanium atom have been investigated with XAFS (X-ray Absorption Fine Structure) for TiO_2 whose diameter is 40, 20, 5, and 3 nm. The results indicate that when the particle size becomes smaller, the proportion of the Ti–OH bond on the surface increases and that the average distance of the Ti–O bond

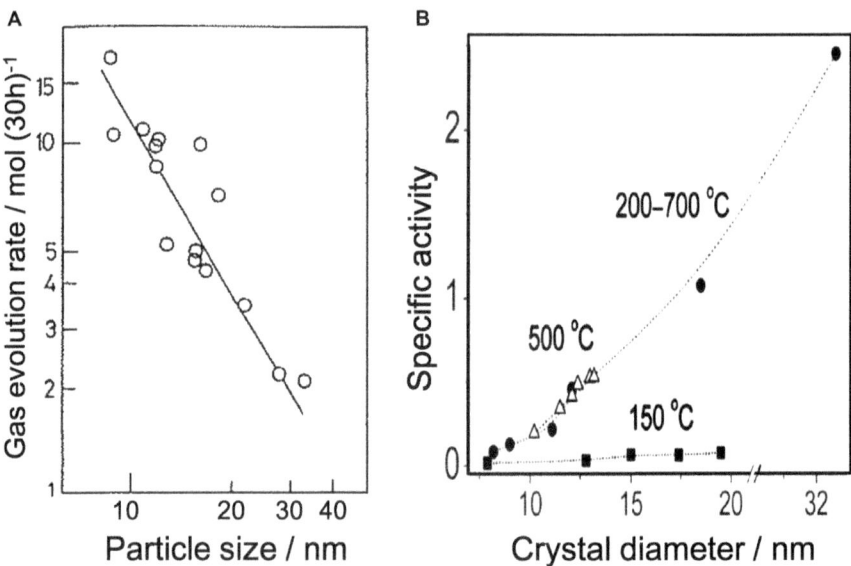

Figure 7.2   Particle size dependence on the photocatalytic reactions, (A) H_2 evolution from methanol solution with deposited Pt and (B) Phenol degradation.
Reprinted (A) from H. Harada and T. Ueda, Photocatalytic activity of ultra-fine rutile in methanol–water solution and dependence of activity on particle size, *Chem. Phys. Lett.*, **106**, 229–231. Copyright (1984), with permission from Elsevier, and (B) from ref. 4 (*J. Phys. Chem. C*). Copyright 2013 American Chemical Society.

becomes shorter, but that the octahedron structure of 6 coordination is retained even for those with particle size of 3 nm.[5] The number of titanium atoms that appear on the surface increases and the crystal edge capable of generating the surface defects increases when the particle size becomes small as deduced from the observation by XAFS.[5] This structural change would bring about the properties characteristic to nanoparticles.

The effects of downsizing of photocatalysts can be summarized as follows. The factors that contribute to the enhancement of the efficiency of the photocatalytic reactions are considered to be:

  (i) Increase in the surface area per weight (or volume);
 (ii) Increase in the number of active sites caused by the increase in the edges of the crystal surface;
(iii) The time to reach the surface is shortened for electrons and holes because photoabsorption takes place near the surface;
(iv) Enhanced oxidation power caused by the valence-band shift owing to the quantum size effect.

To the contrary, the factors to decrease the efficiency of the photocatalytic reactions are considered to be:

  (i) The recombination reaction becomes faster;
 (ii) The sensitivity to light of longer wavelength decreases due to the quantum size effect;
(iii) The generation of a space charge layer for carrier separation becomes difficult.

Thus, there are a lot of factors caused by the downsizing of photocatalysts. However, it is generally believed that the smaller the particle size is, the activity becomes higher although the actual effect might be different for individual catalysts.

7.1.2 Heat Treatment

The outline of the changes caused by the heat treatments of TiO_2 is shown in Figure 7.3. Though the photocatalytic activity is often enhanced by heat treatments,[6] on heating TiO_2 at high temperature, the surface hydroxyl groups are eliminated and water cannot be adsorbed as stated in Chapter 5 (Section 5.1). Many oxygen defects can be generated depending on the condition of the heat treatment. When the defects are generated on the surface, they become

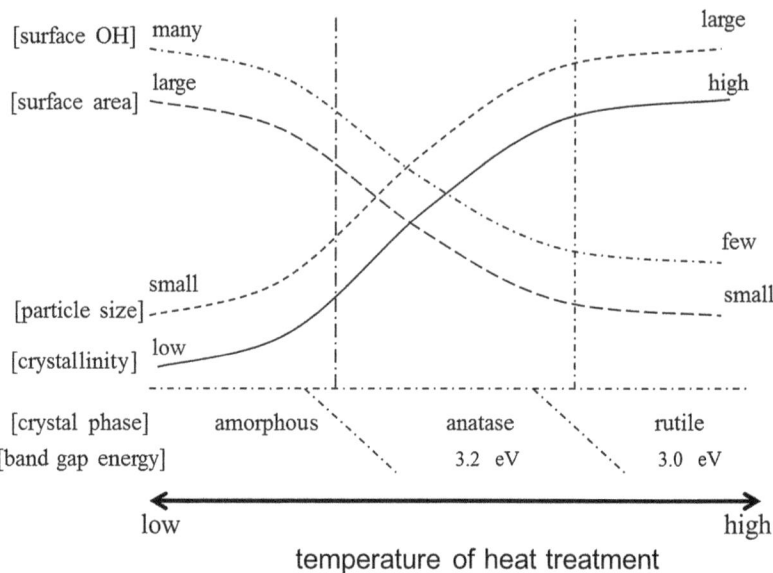

Figure 7.3 Plausible changes in the properties of TiO₂ caused by heat treatments.

reactive sites. On the other hand, when they are generated inside the catalyst, they are considered to act as sites for the recombination.[7]

On heat treatment of hydrous TiO₂ above 400 °C, the formation of anatase crystal can be confirmed by XRD. The surface area decreases while the photocatalytic activity is remarkably enhanced on the crystallization of TiO₂. It is generally believed that the possibility of the recombination decreases by the improvement of the crystallinity, which surpasses the effect of the decrease of the amount of the adsorbed substrates due to the decrease in the surface area. On increasing heat treatment temperature, the anatase crystal grows. Within a certain temperature range, a small difference in the treatment temperature can cause a large change in the crystal size and surface area, which are difficult to be controlled accurately.[8] For any sample, the phase transition from anatase to rutile takes place at around 600 °C. However, when the impurities such as sulfate ions are contained, they prevent the phase transition and crystal growth resulting in a decrease in the surface area.[8]

The excessive heat treatments reduce the photocatalytic activity because the surface area is remarkably decreased and/or the defects are formed inside the crystal. Therefore, several methods to prepare nanoparticles in aqueous solutions by use of various organic compounds are reported.[9] There is a report on the decahedron TiO₂ with a good crystallinity of anatase fine particles, but it is not always the case that the photocatalytic reaction rate is enhanced, even if it is a complete single crystal. It was also reported that the higher the number of surface defects, the higher the activity becomes.[10] Therefore, the surface defects seem to be necessary as the active sites.

7.1.3 Mechanical Downsizing

To prepare nano-particulates of metal oxides, grinding by a mortar or ultrasonication has often been adopted but what happens in this process has not been discussed.

On grinding the mixed crystalline TiO_2 (Degussa P25) at low energy with an agate ball mill, the particle size of TiO_2 of anatase crystallite decreases from 20 to 7 nm after 8 h grinding, while the size of rutile crystallite decreases from 33 to 7 nm after 25 h grinding.[11] By grinding, the anatase phase transfers to the rutile phase. At first, the ratio of rutile crystallite is 20% but after the 20 h grinding it becomes almost 100%, and the amorphous phase becomes only partly observable.

The crystal phase of Hombikat UV-100 (Sachtleben Chemie GmbH) is 100% anatase TiO_2. It becomes completely brookite phase after 150 h grinding and further grinding leads to transformation into the rutile phase.[11] Furthermore, for ST-01 TiO_2 (Ishihara Sangyo Co. Ltd), which is 100% anatase crystal phase the same as UV-100, it was indicated that by treating with a ball mill, the crystal form transferred from anatase → brookite → rutile.[12]

Because the defects inside the crystal generally increase by grinding, for any TiO_2, the photocatalytic activity decreases by one order of magnitude. Especially for P25 the photocatalytic activity is reduced after a short time grinding. Therefore, the technique utilizing grinding to prepare nanoparticles of photocatalysts is not recommended.[11]

7.2 Crystal Dependence of TiO_2

7.2.1 Effect of Crystal Phase

Though several crystalline phases of TiO_2 are known,[13] photocatalytic activity was investigated mainly in anatase and rutile crystals. The amorphous TiO_2 which takes no crystal form does not show photocatalytic activity.[14] Different from crystals, for amorphous TiO_2, unsaturated bonds (non-stoichiometric bonds) are readily formed and the recombination between electrons and holes is considered to increase.

Most of the commercial TiO_2 as photocatalysts are of anatase crystal form. It is said that the anatase TiO_2 shows generally higher photocatalytic activity than rutile TiO_2. The items that affect the photocatalytic activity are the crystal size, the difference in the density of carriers based on the difference of the photoabsorption spectra, and the difference in dispersibility in the case of suspension.

It has been experimentally indicated that the recombination rate between photogenerated electrons and holes is higher for rutile than for anatase.[15] This is ascribed to the fact that indirect bandgap excitation takes place for anatase while direct bandgap excitation takes place for rutile TiO_2 (see Figure 2.6A) and the relatively lower recombination is one of the factors for the high photocatalytic activities for anatase TiO_2. Because the effective mass of holes for the rutile crystal is larger by one order of magnitude than that for the anatase crystal (see Table 1.1), the mobility of the holes is intrinsically small. In other words the frequency of the recombination for rutile crystal is likely to be large because the spread of the holes (effective Bohr radius) is small for rutile and the time to reach the surface is long.

The high activity for anatase has also been interpreted by the fact that the conduction band edge of anatase crystal is higher than that of rutile, which is profitable for the reduction reaction of O_2 which is believed to be the rate determining process for photocatalysis. However, as stated in Chapter 4, it has been recently suggested that the low edge of the conduction band should be lower for anatase TiO_2. Actually, by measuring the production amount of $^{\bullet}O_2^{-}$ it has been reported that the reduction reaction of O_2 is more rapid for rutile crystal.[16] The lower conduction band edge for anatase suggests that the valence band edge calculated from E_g of 3.2 eV should become lower, indicating a higher oxidation potential, which leads to the higher activity.

Besides these factors, the difference in the photocatalytic activities on the various surfaces of TiO_2 would also not be neglected. Namely, as indicated in Chapter 1, the distances between Ti atoms which appear on the first layer are 296 and 379 pm for rutile and anatase, respectively, because the arrangements of the Ti-centered octahedron for rutile and anatase are different. Therefore, as shown in Figure 7.4, the oxidation mechanism is considered to be different.[17] Namely, rutile TiO_2 is capable of forming the surface structure such as Ti–OO–Ti, leading it to readily form O_2. It is well known that the

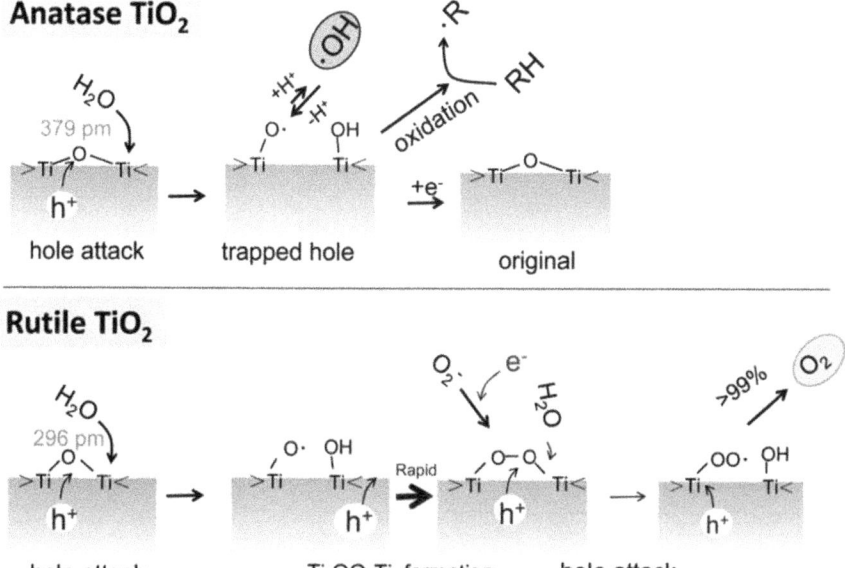

Figure 7.4 Plausible photocatalytic oxidation processes at the surface of TiO$_2$. The oxidation of organic molecule RH for anatase, and that of water for rutile polymorph.
Reproduced from Y. Kakuma, A. Nosaka and Y. Nosaka, *Phys. Chem. Chem. Phys.*, 2015, **17**, 18691, with permission from the PCCP Owner Societies.[16]

activity to generate O$_2$ for rutile is higher than that for anatase TiO$_2$.[18] On the other hand, anatase cannot take the structure of Ti–OO–Ti and the OH radical is considered to be easily produced.[16]

7.2.2 Mixed Crystal Phase

Rutile–anatase mixed phase TiO$_2$ crystallites show interesting photocatalytic properties which are not seen in any single phase.[19] The commercial fine particle TiO$_2$, Degussa P25, is used worldwide as a research material of photocatalyst powder with high photocatalytic activity. The high photocatalytic activity is explained by the fact that it contains both rutile and anatase at the ratio of 2 : 8 and that the rutile and anatase crystallites work cooperatively. The effects have been thoroughly discussed.

The inside structure of the TiO$_2$ particle actually transfers to rutile on successive calcinations, and it was confirmed that the anatase crystal is connected with the rutile crystal as shown by the electron micrograph in Figure 7.5A, and the synergetic effect was observed.[20]

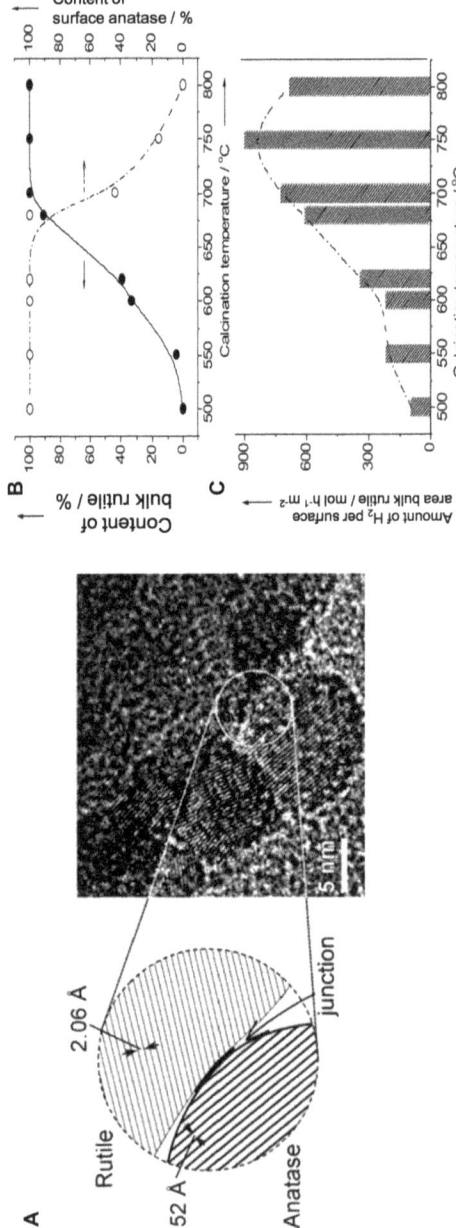

Figure 7.5 (A) HRTEM image of an anatase-rutile mixed phase TiO$_2$ sample. Dependences of the calcination temperature of the samples on (B) bulk rutile content and surface anatase content, and (C) surface-specific photocatalytic activity.

In this system the H_2 generation reaction was studied with TiO_2 photodeposited with platinum using methanol as a sacrificial reagent. The activity became maximum at the state that about half of the surface is covered with anatase as shown in Figure 7.5B and C. When the surface anatase is eliminated, the activity decreases.

Furthermore, for the experiments measuring the decomposition of methylene blue by changing the ratio of anatase to rutile of porous TiO_2 formed on the thin film, it was reported that for 60% anatase the activity was enhanced by 50%.[21]

On the other hand, a lot of reports that such a synergetic effect is doubtful have been presented.[19] Anatase and rutile crystallites were isolated from Degussa P25 by selective dissolution with a hydrogen peroxide–ammonia mixture and diluted hydrofluoric acid, respectively. Comparison of activities of original P25 and reconstructed P25 with those of isolated anatase and rutile particles suggested a less-probable synergetic effect of the co-presence of anatase and rutile.[22] However, one might point out that this result could be attributed to the inefficient connection of rutile and anatase crystallites.

It was indicated, by low temperature ESR measurements for P25 TiO_2, that on shifting the photoexciting wavelength close to visible light wavelength, most of the photons are absorbed at the rutile part, and that the trapped electrons generated in the rutile part transfer to the trapped electrons at the anatase part.[23] Therefore, it is the cause of the high photocatalytic activity of P25 that the rutile crystal actually efficiently absorbs photons like an antenna, and that the charge separation of the electron–hole is enhanced due to the electron transfer from rutile to anatase crystals, leading to the decrease of the recombination rate of electrons with holes.[23]

As stated in Chapter 4 (Section 4.3.2), the flat band potential estimated from the Mott–Schottky plot is higher for anatase, and it was considered that it is hard for the electron transfer to occur from the conduction-band bottom for rutile compared to that for anatase. However, it has been reported that the conduction band of anatase is lower by 0.4 V than that of rutile.[24] The flat band measurements stated in Chapter 4 can be interpreted as that the edge of the conduction band corresponds to the direct transfer with the bandgap energy of 3.8 eV as shown in Figure 7.6A. In the DOS figure shown in Figure 7.6B, fewer states can take the conduction band edge with the presence of indirect bandgap. Therefore, the transfer of excited electrons from rutile to anatase crystals is considered to occur more easily because the conduction band bottom for anatase of indirect bandgap is lower than that for rutile of direct bandgap.

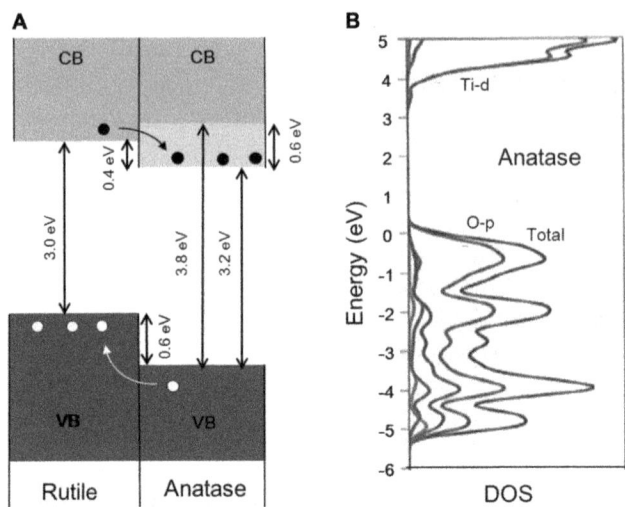

Figure 7.6   (A) Suggested band structure of mixed phase TiO_2 modified to distinguish indirect (3.2 eV) and direct (3.8 eV) bandgaps for anatase and (B) the calculated density of states of anatase TiO_2. Reprinted by permission from Macmillan Publishers Ltd, D. O. Scanlon, C. W. Dunnill, J. Buckeridge, S. A. Shevlin, A. J. Logsdail and S. M. Woodley, Band alignment of rutile and anatase TiO_2, *Nat. Mater.*, **12**, 798. Copyright (2013).

7.2.3   Effect of Crystal Facet

As was comprehensively reviewed, the photocatalytic activity depends on the crystal facet.[25] To investigate the crystal facet dependency a queue in the different crystalline direction was prepared, and then the reductive-deposition of Ag was examined.[26] The facet dependency observed is not consistent with that for a single crystalline particle.[27]

In Figure 7.7A, the features of the precipitation of platinum by the reductive photocatalytic reaction are shown in (a) and (b) for anatase and rutile crystallites, respectively. In (c) and (d), deposition of PbO_2 by the oxidative photocatalytic reaction of Pb^{2+} on the anatase and rutile crystallites are shown, respectively. As illustrated in Figure 7.7B, for the rutile surface in the suspension system, (110) works as a reductive site and (101) functions as an oxidative site. For anatase the (001) facet is more involved in the oxidative reactions than the (101) facet.[27] Furthermore, for anatase, when a tetradecahedron of 14 faces was formed from a decahedron of 10 faces by generating (100) facet, oxidation took place at this facet and the photocatalytic efficiency was further enhanced.[28] As suggested in the energy band diagrams of both crystals (Figure 3.10), the energy levels for the valence band top and

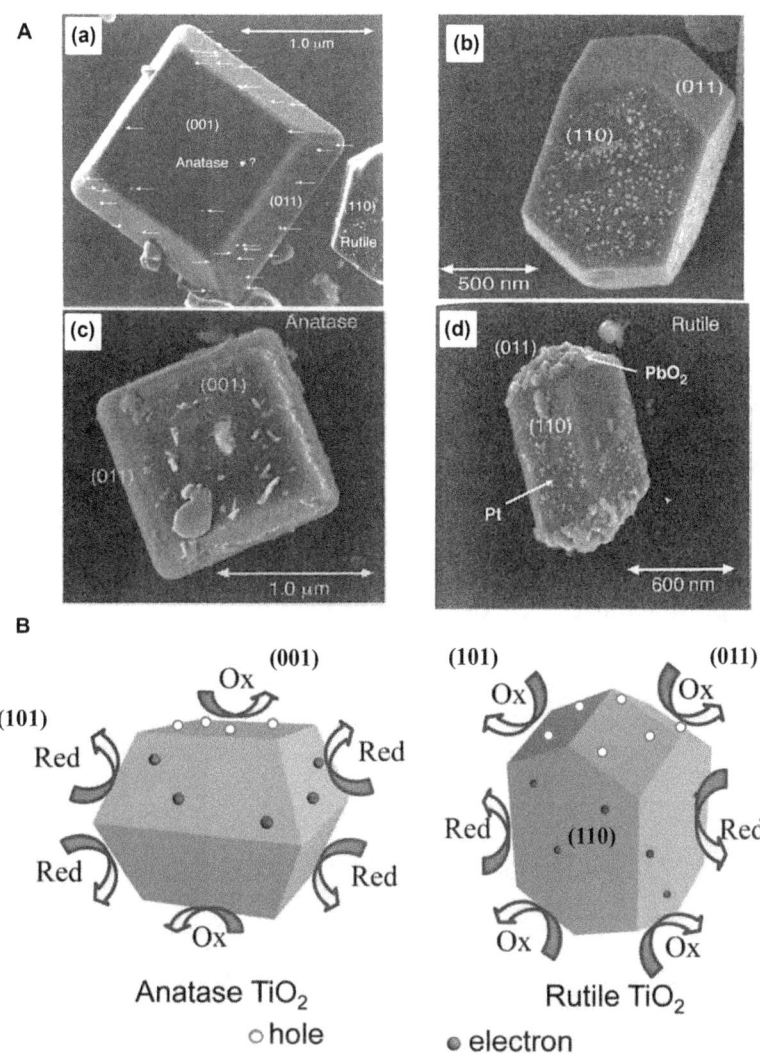

Figure 7.7 (A) SEM images of anatase (a,c) and rutile (b,d) particles on which Pt particles were deposited (a,b) and PbO$_2$ particles were deposited (c,d) by further photocatalytic reaction. For anatase, (011) = (101). Reproduced from ref. 27 (*New J. Chem.*, 2002) with permission from CNRS and RSC. (B) Schematic representation of the spatial separation of redox sites on an anatase and a rutile TiO$_2$ crystallite. Reproduced with permission from R. G. Liu, G. Q. Lu, H.-M. Cheng *et al.*, *Chem. Rev.*, 2014, **114**, 9559. Copyright 2014 American Chemical Society.[25]

the conduction band bottom are different depending on the difference of the crystal facets. Therefore, when holes and electrons move in the single crystals, the direction of the movement can be different against the crystal axes. For instance, it can happen that holes appear more favorably at (001) and (100) facets than the others.

7.3 Doping of TiO₂ Powder

The introduction of substances such as ions into the inside of a crystal is called doping. By doping, the electronic structure of TiO_2 can be altered by creating an intermediate state in the bandgap and/or narrowing the gap itself. Figure 7.8 summarizes the relative positions obtained by DOS calculation for various doping elements to the band edges of intrinsic TiO_2.[29] It has been gathering attention because the absorption of visible light becomes possible by doping to semiconductors which are unable to use the energy of light other than UV for photocatalytic reactions.

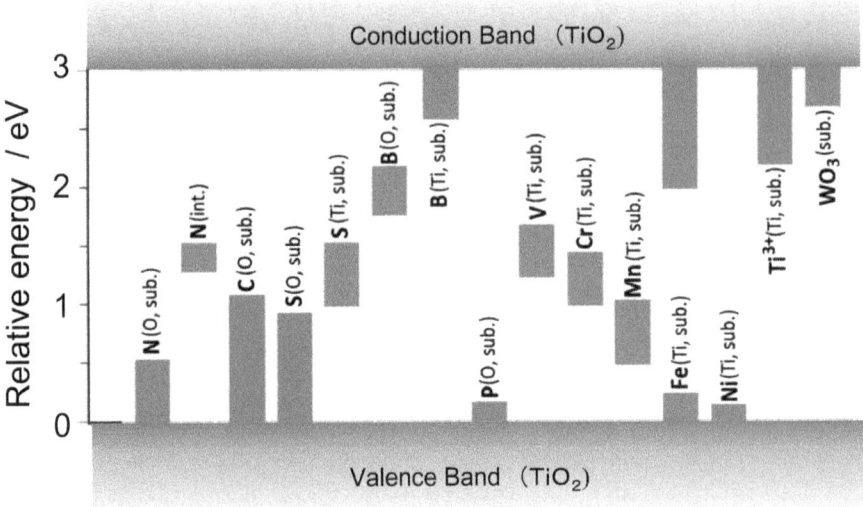

Figure 7.8 Schematic illustration of energy level positions for various dopants in TiO₂ relative to band edges.
Reproduced from I. Paramasivam, H. Jha, N. Liu and P. Schmuki, A review of photocatalysis using self-organized TiO₂ nanotubes and other ordered oxide nanostructures, *Small*, 2012, **8**, 3073–3103. Copyright © 2012 WILEY-VCH Verlag GmbH & Co. KGaA, Weinheim.

7.3.1 Nitrogen Doping

There are two kinds of nitrogen doping. Doping by oxygen-substitute nitrogen, N(O, sub.), causes narrowing of the bandgap by introducing N2p state just above the TiO_2 valence band as shown in Figure 7.8. In the second doping, an intermediate state is formed almost in the middle of the bandgap due to interstitial nitrogen, N(int.). The former doping is effective for visible light photocatalytic activity and, as shown in Figure 7.9A, the reductive ability is considered to be the same as that of pure TiO_2 although the oxidative ability is more or less decreased. In Figure 7.9, energy levels of metal ions for doping and deposition, which will be described later, are shown together.

TiO_2 doped with nitrogen, N–TiO_2, was found to actually photodecompose acetaldehyde under visible light irradiation.[30] However, oxygen defects in the crystallite are generated, which can shorten the lifetime of the photoinduced conduction band electrons to keep the

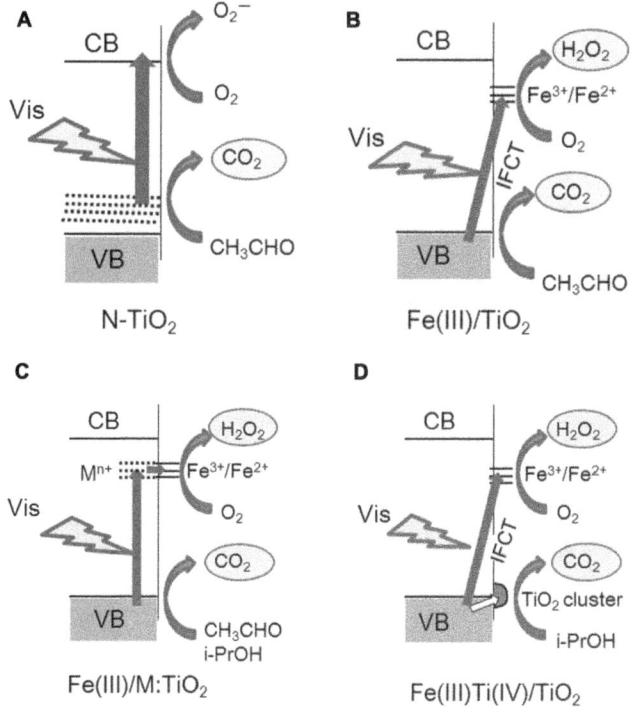

Figure 7.9 Schematic energy diagram for (A) nitrogen doped TiO_2, (B) Fe(III) deposited TiO_2, (C) Fe(III) deposited metal-ion doped TiO_2, and (D) Fe(III) and Ti(IV) deposited TiO_2. IFCT: interfacial charge transfer.

electronic charge because the oxidation number for O is -2 while that for N is -3.[31] Then, various treatments to enhance the activity have been attempted for N–TiO$_2$.[32] In Figure 7.10, examples of the enhancement of the photocatalytic activities by depositing metal ion compounds on the surface are shown.[33]

Figure 7.10A shows the relationships between the redox potentials of deposited various metal species and the photocatalytic activities of the modified N–TiO$_2$ in aqueous suspension, which were tested for the decomposition of various VOC (Volatile Organic Compounds). The results in this figure suggest that the activity of N–TiO$_2$ should be remarkably enhanced by the addition of metal species such as V$^{5+/4+}$, Fe$^{3+/2+}$, Cu$^{2+/+}$, and Pt$^{4+/3+}$, whose redox potentials are in the regions of about $+0.6$ to $+1.0$ V $vs.$ SHE.[33] Figure 7.10B shows the dependence of CO$_2$ yields on the photoirradiation wavelength for N–TiO$_2$ and VCl$_3$ deposited N–TiO$_2$ (VCl$_3$/N–TiO$_2$). As shown in Figure 7.10B, for VCl$_3$/N–TiO$_2$, the responsive wavelength for the CO$_2$ yield expands to a longer wavelength (b) than that for N–TiO$_2$ (a). The deposition of metal ions will be described later in Section 7.4.1.

Besides N, non-metal dopants such as C, F, S, and B are known to modify TiO$_2$ photocatalysts.[34]

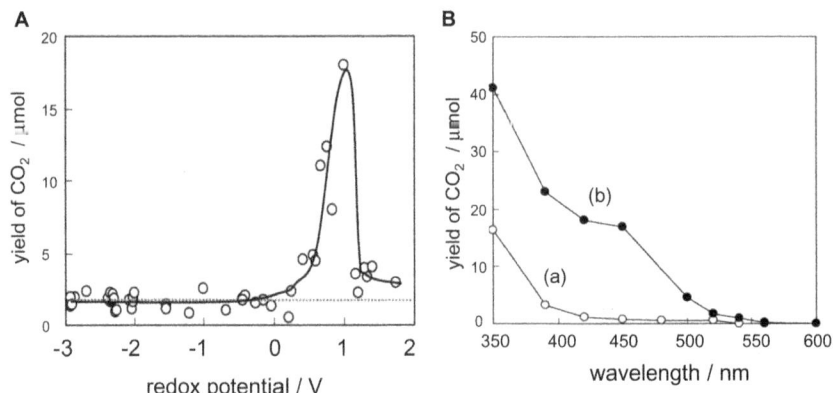

Figure 7.10   (A) Relationship between the redox potentials of metal species deposited and the photocatalytic activities of N–TiO$_2$. The dotted line represents the activity of N–TiO$_2$. (B) Dependence of CO$_2$ yields on the photo-irradiation through different low cut-off filters over (a) N–TiO$_2$ and (b) VCl$_3$/N–TiO$_2$.
Reprinted from S. Higashimoto, W. Tanihata, Y. Nakagawa, M. Azuma, H. Ohue and Y. Sakata, Effective photocatalytic decomposition of VOC under visible-light irradiation on N-doped TiO$_2$ modified by vanadium species, *Appl. Catal., A,* **340**, 98–104. Copyright (2008), with permission from Elsevier.

7.3.2 Metal Ion Doping

When metal oxides such as TiO_2 are prepared with a sol–gel method, metal ions can be readily doped. On homogeneous doping of transition metal ions such as Cr^{3+}, photoabsorption in the visible light regions takes place. This kind of research was performed intensively in the 1980s. However, when the energy level of the metal ions doped inside the particle is involved in the photoexcitation, the photogenerated carriers are trapped at the metal ions and cannot move to the surface. Therefore, the carriers are not used for the photocatalytic reactions but the recombination is considered to be accelerated. Then it was concluded that the photocatalytic activity substantially decreased due to the accelerated recombination of electrons and holes.[35]

The efficiency depends on the method of preparation. For instance, it has been reported that on doping rare earth metals to TiO_2 not only by visible light but also by UV light the activities to decompose dye were enhanced.[36] When TiO_2 and $SrTiO_3$ are doped with Cr^{3+}, the recombination increases because when the location of tetravalent Ti^{4+} is substituted by Cr^{3+} the oxidative state of Cr^{3+} becomes unstable.[37] Then, by co-doping stable pentavalent Sb^{5+} or Ta^{5+} ions with Cr^{3+}, the electric charge was considered to be compensated to prevent the formation of the recombination site. Actually, the photocatalysts possessing the oxidative activity under the visible light irradiation have been successfully fabricated by co-doping 1.25% Sb^{5+} and 0.5% Cr^{3+}, in TiO_2 and $SrTiO_3$.[37]

As stated above, doping of metal ions inside the particle is apt to promote the recombination of photoinduced electron–hole pairs. Hence, it would be effective to dope the metal ions to only the part close to the crystallite surface, which are adopted in the recent literature on doping.[38]

7.4 Surface Modification

7.4.1 Metal Ion Deposition (Grafting)

As briefly stated in Section 7.3.1, the deposition of various co-catalysts is one of the modification methods to enhance photocatalytic activity for redox reactions. Binding a small amount of compounds on the surface, but not inside the crystallite, is called deposition, and in some cases it is called grafting.

The metal ions become clusters of hydroxides or oxides depending on the methods to deposit on the surface and are loaded as shown in

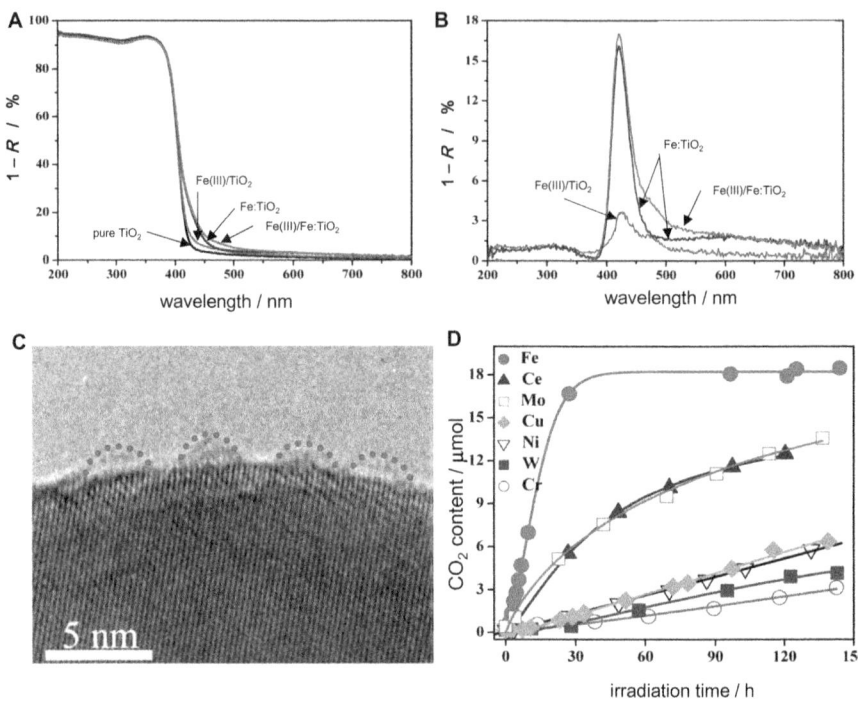

Figure 7.11 (A) UV-vis reflectance spectra of TiO_2, Fe(III) deposited TiO_2 (Fe(III)/TiO_2), Fe^{3+} doped TiO_2 (Fe:TiO_2), and Fe(III) deposited Fe^{3+} doped TiO_2 (Fe(III)/Fe:TiO_2). (B) Difference spectra from pure TiO_2. (C) HRTEM images of Fe(III)/Fe:TiO_2. (D) CO_2 generation curves for Fe(III)/M:TiO_2 (M = Fe, Ce, Cu, and Ni) samples under visible-light irradiation with iso-propanol.
Reprinted with permission from M. Liu, X. Qiu, M. Miyauchi and K. Hashimoto, *J. Am. Chem. Soc.*, 2013, **135**, 10064. Copyright 2013 American Chemical Society.[38]

the electron micrograph in Figure 7.11C.[38] For photocatalysts used for environmental clean-up, enhancement of the activity can be expected by the deposition of Cu^{2+} or Fe^{3+}, which become reduction catalysts for O_2 because the species to be reduced are always O_2. The excitation from the valence band to the deposited ions (see Figure 7.9B), *i.e.*, interfacial charge transfer (IFCT) absorption, is observed as shown in Figure 7.11A and B although the absorbance is small. On the excitation of the IFCT band, the absorption energy is used for electron transfer to the co-catalysts without loss. This is more efficient than the photoinduced conduction band electrons in the semiconductor moving to the surface of the co-catalysts because in the latter case recombination with holes could occur during the movement.[39]

On doping the metal ions stated in the former section, the absorbance substantially increases as compared to the IFCT absorption. Though the photocatalytic activity increases by doping various metal ions, the activity enhances best (see Figure 7.11D) by doping the same Fe^{3+} as the deposited metal ions.[38] This would be ascribed to the smooth transfer of electrons from the dope level to the deposition level as shown in Figure 2.9C.

The Ti(IV) cluster that has the same component as TiO_2 works as oxidative catalysts when it is deposited on a TiO_2 photocatalyst as shown in Figure 7.9D, and the activity is reported to be enhanced.[40]

7.4.2 Fluorination of TiO_2

Fluorination of semiconductor photocatalysts can be employed to modify their surfaces and bulk properties, and consequently to enhance their photocatalytic performance.[41]

Because hydrofluoric acid is a strong acid, it is substituted for the surface hydroxyl groups on the TiO_2 surface. The substitution can be confirmed by the fact that no charge transfer absorption band is observed by adding resorcinol which causes the charge transfer band in the absorption spectrum by adsorbing on the surface.[41] The adsorbed amount of the hydrophobic molecule such as phenol increases to accelerate the photocatalytic decomposition because the surface of TiO_2 becomes hydrophobic on adding NaF as shown in Figure 7.12A. The effect depends on pH as shown in Figure 7.12B, which is in good accordance with the pH dependency of Ti–F bond formation on the surface.[42]

However, such rate enhancement is operative with anatase but not with rutile.[43] As a modification, the fluoride ions in both the inner and outer Helmholtz layers can contribute to the desorption of the surface-bound OH radicals and thus to the production of mobile free hydroxyl radicals, due to the mediating effect of fluorine hydrogen bond as shown in Figure 7.12C.[41] Figure 7.12D shows how photocatalytic activity is enhanced; the surface Ti–F group can act as a site to tightly trap the photogenerated electrons due to the strong electronegativity of the fluorine and then to transfer them to O_2 adsorbed on the surface of TiO_2.[44]

7.4.3 Phosphate Treatment of TiO_2

Phosphate anions are known to strongly adsorb onto the surfaces of TiO_2 to form phosphate modified TiO_2 (P-TiO_2). A small

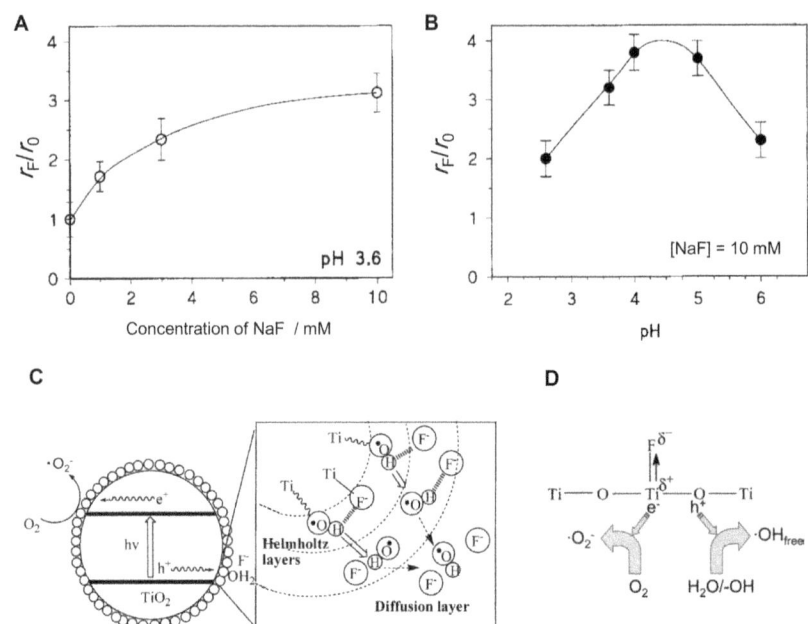

Figure 7.12 (A) Ratio of phenol degradation rates r_F/r_0 in the presence of different concentrations of NaF to that in the absence of NaF at pH 3.6, and (B) the pH dependence with 0.01 M NaF[42] (C). Illustration showing that fluoride ions in both the inner and outer Helmholtz layers contribute to the release of surface-bound hydroxyl radicals through a fluorine hydrogen bond, producing mobile free hydroxyl radicals. (D) Schematic diagram for generation and transfer of charge carriers in F-TiO$_2$ under UV irradiation.
Reprinted from (A), (B) C. Minero, E. Pelizzetti *et al.*, *Langmuir*, 2000, **16**, 2632,[42] (C) Y. Xu *et al.*, *J. Phys. Chem. C*, 2007, **111**, 19024,[43] and (D) J. G. Yu *et al.*, *J. Phys. Chem. C*, 2009, **113**, 6743[44] with permission from American Chemical Society.

amount of phosphate of the catalysts enhances the photocatalytic activity measured by phenol degradation.[45] However, a further increase of the P/Ti molar ratio (>0.01) led to a substantial loss in activity. The optimal calcination temperature of P-TiO$_2$ was 300–500 °C.[45]

Charge carrier dynamics was investigated for P-TiO$_2$ films by transient absorption spectroscopy.[46] The photogenerated charges of the modified film have a much longer lifetime than those of the un-modified film. These differences are attributed to the surface-carried negative charges of TiO$_2$ resulting from the phosphate groups (–Ti–O–P–O$^-$).[46]

7.5   Composite with Semiconductors

In general, on combining a semiconductor with the other semi-conductor or metal, a barrier is formed as shown in Figure 7.13A(b) or B(b). Namely, on combining the semiconductors or metals as illustrated in (a), the band shifts like (b) to reach the equilibrium so as to coincide with the Fermi level E_F. For n-type semiconductors, the difference between E_F and E_C, which is the energy of the conduction band bottom, is determined by the carrier density, N_0, as indicated by eqn (4.13). Because of the difficulty to estimate the carrier density for the particles, the position of E_F is not certain, and in most cases E_C is assumed to be close to E_F as discussed in Chapter 4.

For general connection (b), when E_F shifts the energy position at the interface is retained and the band energy E_C in the semiconductor shifts along with the appearance of the space charges around the interface. As discussed already in the chapter on photoelectrochemistry

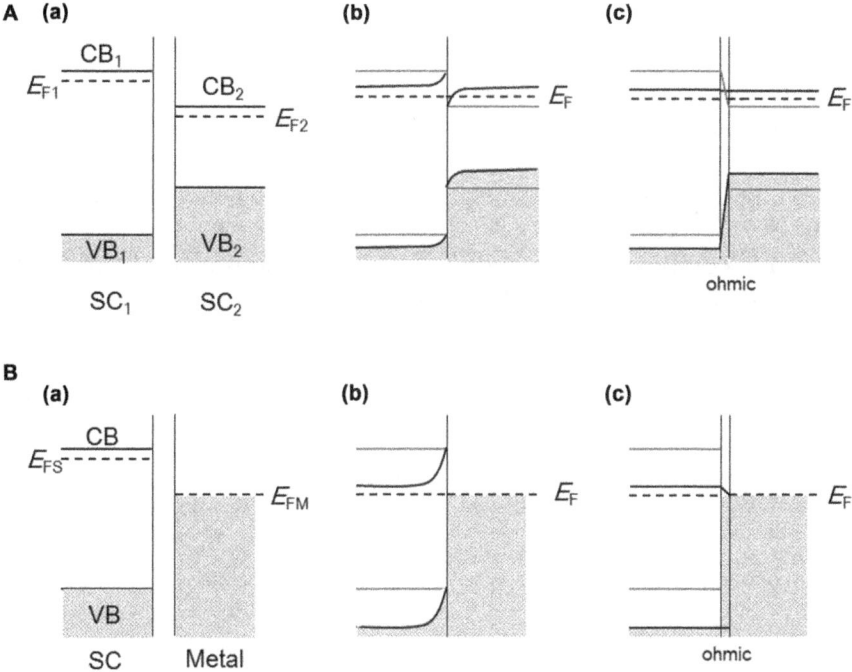

Figure 7.13   Schematic illustration of energy level shift at the interface of n-type semiconductor-1 by contact with (A) n-type semiconductor-2 and (B) metal. (a) Before the contact, (b) after normal contact, and (c) after ohmic contact. Gray lines indicate the band edges before the contact.

(Chapter 4), for the solution containing electrolytes at a high concentration, space charge layers would be formed the same as metals as in Figure 7.13B(b).

When semiconductor particles are deposited on different semi-conductor particles in the photocatalytic systems, the positions E_{F1} and E_{F2} shift to the common E_F which is determined by the amount of carriers in each semiconductor. Therefore, different from the case of the metal deposition, E_F would not change by depositing a small amount of semiconductor nano-particles provided that the carrier densities of each semiconductor are similar.

There is a case that the barrier is not formed at the interface as shown in Figure 7.13(c). Such an interface mode is called ohmic contact. The ohmic contact is generated by placing a thin metal film at the interface, or by the change of the physical properties due to the diffusion of ions by heat treatments. In the case of ohmic contact, E_C is not retained at the interface but moves along with E_F. In the case of the deposition of noble metals, such a contact would be generated by heat treatment at a higher temperature.

As shown in Chapter 1 (Figure 1.8 (c)), a combination of n- and p-semiconductors with ohmic contact would be effective for the reduction and the oxidation of substances. The reaction in which the oxidation and the reduction are performed on each semiconductor separately (Figure 7.14A) is called a Z-scheme. Application to the decomposition of water by sunlight is trying, as will be described in Chapter 10 (Section 10.1.2).

For this composite semiconductor, the electron transfer from CB_1 to VB_2 should be rapid and the holes generated at VB_1 should not move to VB_2 but be consumed by the reactions. Because VB_2 is fully occupied with electrons, at first SC_2 is excited followed by the reduction reaction then the holes generated at VB_2 wait for recombination with the

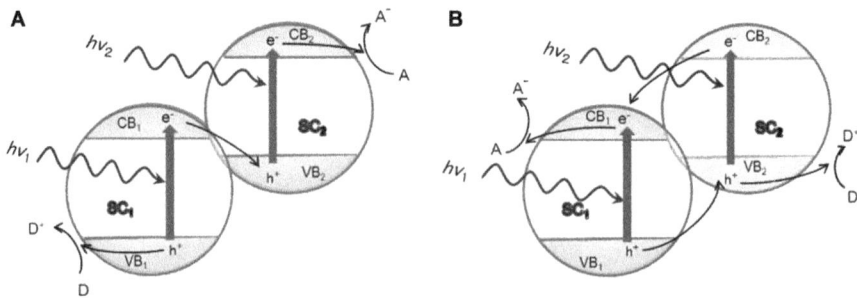

Figure 7.14 Schematic energy diagram for (A) ideal and (B) ineffective paths for a composite photocatalyst.

electrons at CB_1 generated by the excitation of SC_1. This procedure is considered to enhance the reaction efficiency. For the oxidation of water, until SC_1 and SC_2 accept four photons the reaction is not completed. Then, it is not easy to fabricate the desired composite. In many researches, sacrificed reagents are used for A or D to achieve the reaction pass shown in Figure 7.14A.[47] The charge transfer shown in Figure 7.14B occurs comparably more easily as was seen in Section 7.2.2 for the anatase–rutile composite semiconductor (see Figure 7.6A). It is natural because part of the bandgap of each semiconductor is retained as the common part of the gap for both semiconductors. This case is ineffective sometimes, because the oxidation and reduction ability of the original semiconductors are both degraded.

7.6 Deposition of Noble Metals

7.6.1 Plasmonic Photosensitizer

Gold (Au) nanoparticles stabilized by citric acid can be deposited on TiO_2 powder, followed by heat treatment at rather low temperature, *e.g.*, 200 °C. After washing, Au deposited TiO_2 (Au/TiO_2) can be fabricated. As indicated previously, physisorbed water on TiO_2 is eliminated at 200 °C, which is low for the diffusion of metal ions toward the surface of TiO_2. Therefore, the barrier of the band edge is formed as shown in Figure 7.13B(b).

Gold nanoparticle presents a broad absorption around 550 nm called SPR (Surface Plasmon Resonance). With the light energy absorption, electrons in Au transfer to the conduction band of TiO_2 as shown in Figure 7.15A.[48] The electrons reduce the oxygen in the air to generate $^\bullet O_2^-$, while the holes remaining in the Au nanoparticles are involved in the oxidation reactions. The plasmon-induced photocatalysis worked by such a mechanism gathers attention. Figure 7.15B shows the dependency of the generation rate of $^\bullet O_2^-$ on the particle size of each crystal.[49] Because the conduction band bottom is higher for rutile, the Schottky barrier at the interface of Au also becomes higher, resulting in the lower rate of electron transfer from Au to TiO_2. Furthermore, because the recombination is decreased with an increase of the particle size, the generation rate of $^\bullet O_2^-$ is higher for larger particle size.[49] Au/TiO_2 may be useful because the charge transfer takes place by visible light. However, useful reactions are limited because the oxidative ability of the holes remaining in the Au nanoparticles is not high.[50]

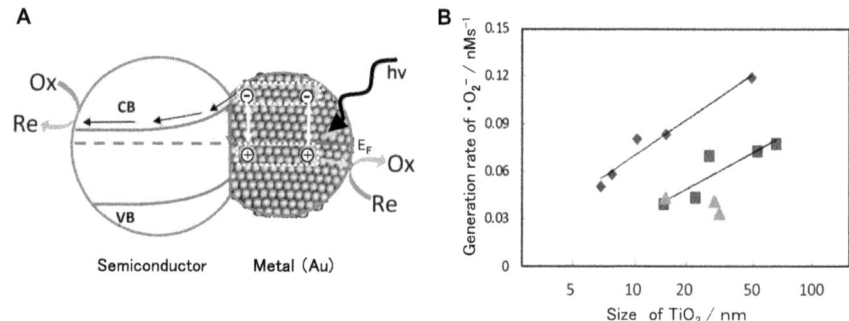

A

B

Figure 7.15 (A) Schematic diagram of plasmon-induced charge separation and associated photochemistry at the metal/semiconductor heterojunction. (B) Initial rate of $^\bullet O_2{}^-$ generation on the 532 nm irradiation as a function of primary particle size of the TiO$_2$ for Au nano-particle deposited TiO$_2$ with different crystalline phases; anatase ($\blacklozenge$), rutile ($\blacksquare$), and mixed phase ($\blacktriangle$). Reprinted with permission from (A) Z. Zhang and J. T. Yates, Jr. *Chem. Rev.*, 2012, **112**, 5520. Copyright 2012,[48] and (B) H. Saito and Y. Nosaka, *J. Phys. Chem. C*, 2014, **118**, 15656. Copyright 2014,[49] American Chemical Society.

7.6.2 Modification of Band Structure

Because the carrier density (the density of mobile electrons) of metals is high, different from semiconductors, it is effective to enhance the photocatalytic activity by depositing a small amount of metals on the semiconductor. For particulate photocatalysts, as shown in Figure 7.16A, the metal–solution and semiconductor–solution interfaces, which correspond to Figure 7.13B (b) and (c), appear on the same surface. Therefore, a two-dimensional potential map must be taken into the account to show the shift of the bands. Under the assumption that the metal and the semiconductor are in ohmic contact and that the Fermi levels coincide, the contour map of the potential was calculated as shown in Figure 7.16B, where the distance between the metal particles of 20 nm, carrier density $N_0 = 10^{18}$ cm^{-3}, and dielectric constant $\varepsilon = 80$ were assumed. The space charge layer generated inside the semiconductor is divided around the surface and a high electric field gradient is generated.[51]

The activity becomes higher with the increase of the work function of deposit metal on plotting the catalytic activity against the work function of metals because the potential difference V_0 corresponds to the difference between the conduction band bottom (for n-semiconductor) and the work function of the deposited metal.[52] To the contrary, in the case of GaP of p-type semiconductor, the activity

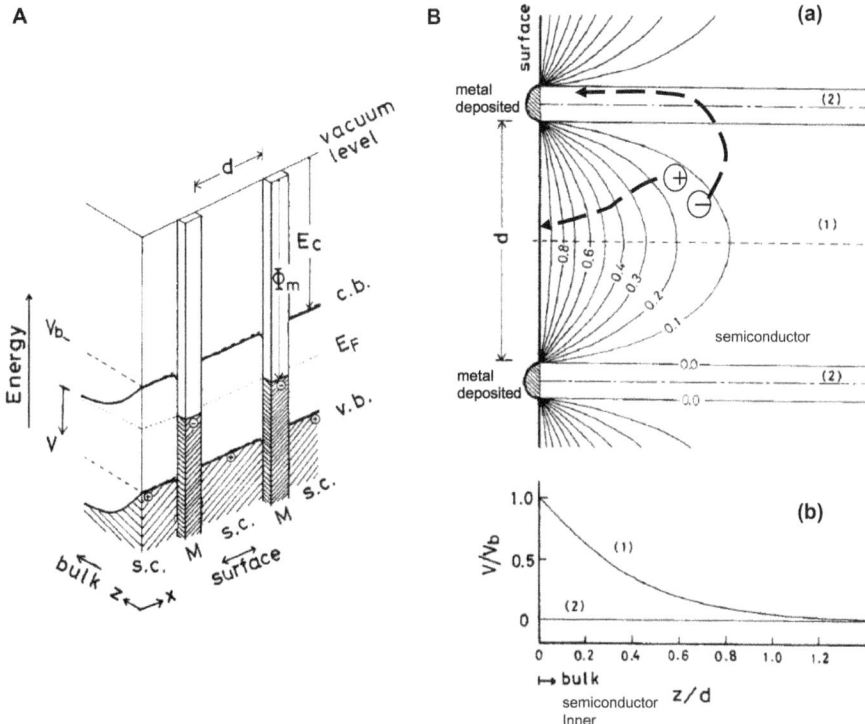

Figure 7.16 (A) Energy level diagram of metal-deposited semiconductor grain. M; metal, S.C.; n-type semiconductor, Φ_M; work function of metal, E_F; Fermi level. (B) Potential energy surface obtained by numerical calculation for Poison–Boltzmann equation. Curves (1) and (2) in (b) are the cross sections along the broken line and the dotted broken line indicated in (a), respectively. Reproduced from Y. Nosaka, Y. Ishizuka and H. Miyama, *Ber. Bunsen-Ges. Phys. Chem.*, 1986, **90**, 1199, with permission from The German Bunsen Society and Wiley-VCH Verlag GmbH & Co.[51]

becomes higher with the decrease of the work function of deposit metal.[52] This indicates that the charge separation is enhanced by the deposition of noble metal.

7.7 Composite with Adsorbents

By depositing TiO_2 on the adsorbent materials on which reactant molecules can easily adsorb, the molecules are concentrated by the adsorbent around TiO_2, resulting in the enhancement of the activity. The correlation of the adsorbability of the adsorbent with the photocatalytic decomposition rate has been investigated by using various adsorbents as composite materials for TiO_2.[53]

On depositing TiO_2 at different ratios on mordenite, which is a kind of cray mineral with high adsorbability, (a) the amount of the adsorbed reactant molecule, propionaldehyde, S_{ads}, and (b) the rate constant for CO_2 generation by the photocatalytic decomposition, k_{CO_2}, are shown in Figure 7.17A. With increasing the amount of TiO_2, the photocatalytic activity becomes higher, but when the amount surpasses 50%, the adsorption amount starts to decrease; accordingly, the photocatalytic activity begins to become lower. This fact indicates that even if the ratio of TiO_2 is increased, the decomposition rate becomes lower when the amount of the adsorption decreases.

For several kinds of adsorbents with 50% TiO_2, (a) S_{ads} and (b) k_{CO_2} were measured and plotted in Figure 7.17B against adsorption constant K_{ad} for each adsorbent. As indicated in Figure 7.17B (a), the adsorption constant for adsorbent i (active carbon) is the largest but the amount of the adsorption S_{ads} is almost the same as

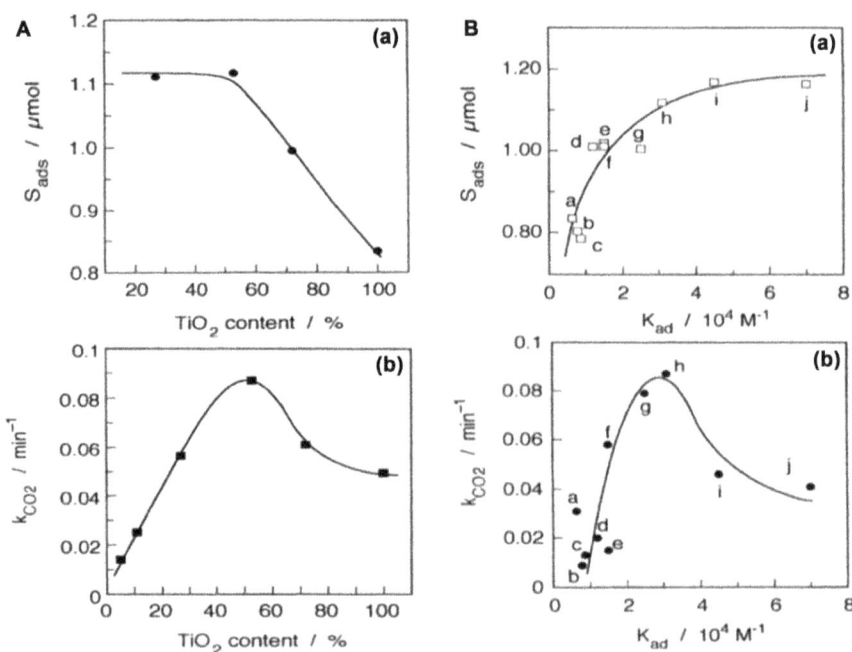

Figure 7.17 (a) The amount of adsorbed propionaldehyde on the photocatalyst films and (b) the CO_2 evolution rate in the photocatalytic decomposition of propionaldehyde as a function of (A) the fraction of TiO_2-loadings on mordenite and (B) adsorption constant K_{ad}. Reproduced from H. Yoneyama and T. Torimoto, Titanium dioxide/adsorbent hybrid photocatalysts for photodestruction of organic substances of dilute concentrations, *Catal. Today*, **58**, 133–140. Copyright (2000), with permission from Elsevier.

adsorbent h (mordenite) whose adsorption constant is medium. On the other hand, as shown in Figure 7.17B (b), the generation rate of CO_2 is higher for h (mordenite). The large adsorption constant means that the adsorbability is strong. However, if the amount of the adsorption is the same, the reaction proceeds more rapidly for the weaker adsorption. Namely, when adsorbability is strong, the rate of the reactant species to diffuse from the surface of the adsorbent to TiO_2 becomes low. Therefore, the generation rate of CO_2 decreases.[53]

7.8 Effect of Additives in Solution

7.8.1 Addition of H_2O_2

The addition of a small amount of hydrogen peroxide (H_2O_2) enhances photocatalytic reactions, but the addition of an excess amount to the contrary disturbs the reactions. When there is no reactant except water and oxygen, the generation of $^{\bullet}O_2^{-}$ increases with an increase of H_2O_2, and that of OH radical also increases for TiO_2 like P25 of mixed crystal containing rutile.[17] To the contrary, for 100% anatase crystal, OH radical generation is decreased with a small amount of H_2O_2.[17] As shown by the thickness of the arrows in Figure 7.18, the generation of OH radical by the oxidation of water decreases due to the oxidation of H_2O_2 for anatase. In the case of rutile, H_2O_2 could act as a catalyst to generate OH radicals by the oxidation of water because the surface structure is suitable for O_2 generation.[16] When the concentration of H_2O_2 is high, the adsorbed molecules completely cover the surface. Thus, the adsorption of O_2 is prevented to decrease the reduction of O_2 and/or H_2O_2 is reduced to H_2O, resulting in the deterioration of the photocatalytic reactions.

H_2O_2 adsorbs on the surface of both anatase and rutile TiO_2 to generate titanium peroxide (Ti–OO–Ti). On the addition of a substantial amount of H_2O_2, TiO_2 becomes yellow and responsive to visible light. On the basis of IR absorption measurement, it is also reported that H_2O_2 coordinates to Ti by binuclear coordination (Ti-η^2-peroxide, $Ti < O_2$) and the reaction proceeds.[54] For instance, it is known that for the photocatalytic oxidation of olefin, epoxide is formed on the addition of H_2O_2.[54]

7.8.2 Addition of Ozone

Ozone (O_3) is practically used for water treatments owing to its strong oxidizability. It absorbs UV light of wavelength less than

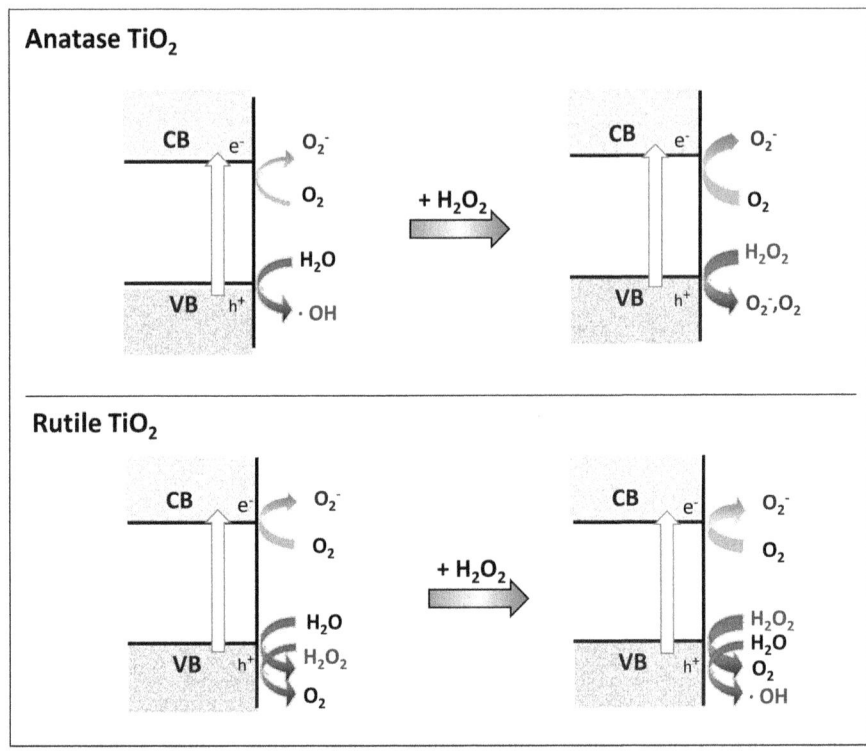

Figure 7.18   Effect of a small amount of H_2O_2 on the photocatalytic processes at the TiO_2 surface of two polymorphs (anatase and rutile) in the absence of organic reactants. The thickness of the arrows expresses the degree of the reaction rate.
Reprinted with permission from J. Zhang and Y. Nosaka, *J. Phys. Chem. C*, 2014, **118**, 10824. Copyright 2014 American Chemical Society.[17]

300 nm and reacts with water to generate several active species such as OH radical. However, because it does not substantially absorb 300–400 nm light, which is used for the excitation of TiO_2, the direct decomposition does not usually take place on the illumination of sunlight and blacklight ($\lambda = 365 \pm 25$ nm). On illumination with blacklight in the presence of TiO_2 photocatalyst, ozone of high concentration (>120 ppm) photocatalytically decomposes by a zero-order reaction and at low concentration it decomposes according to the reaction kinetics of Langmuir–Hinshelwood type.[55] The photocatalytic ozonization by monochloroacetic acid and pyridine proceeds at 4 and 18 times higher rates than those without ozone as shown in Figure 7.19A and B, respectively.[56]

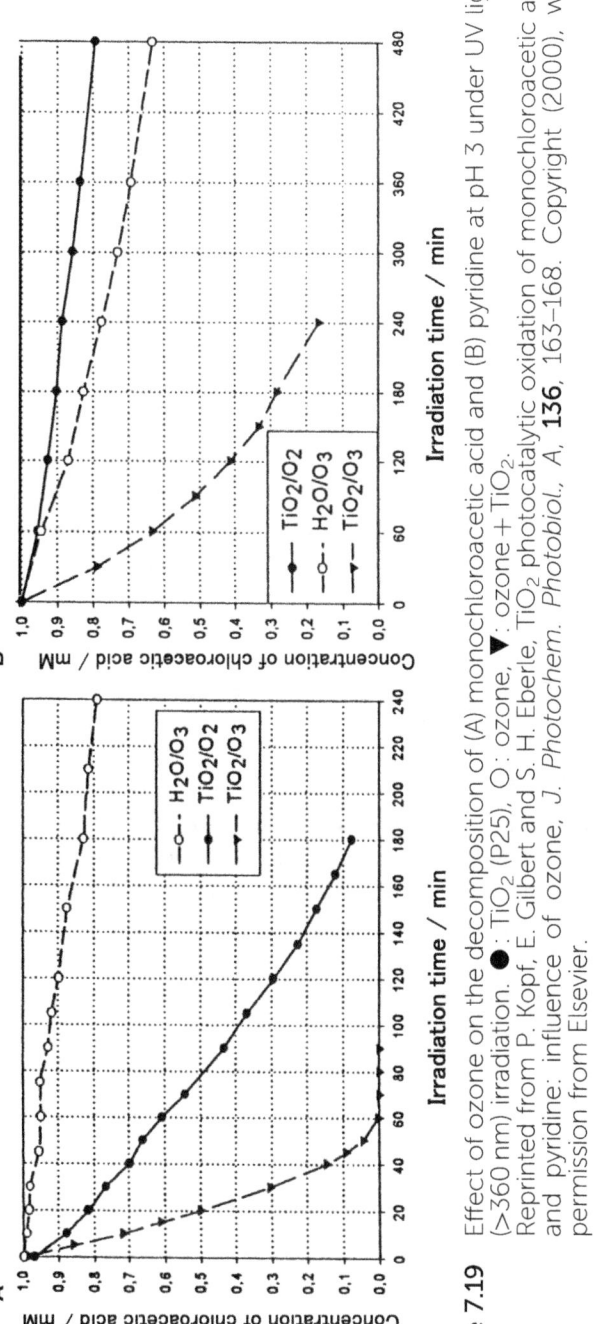

Figure 7.19 Effect of ozone on the decomposition of (A) monochloroacetic acid and and (B) pyridine at pH 3 under UV light (>360 nm) irradiation. ●: TiO_2 (P25), ○: ozone, ▼: ozone + TiO_2. Reprinted from P. Kopf, E. Gilbert and S. H. Eberle, TiO_2 photocatalytic oxidation of monochloroacetic acid and pyridine: influence of ozone, *J. Photochem. Photobiol., A*, **136**, 163–168. Copyright (2000), with permission from Elsevier.

When chlorophenol is adopted as a harmful organic model compound in combined use of ozone in the TiO_2 photocatalytic reaction system, the decomposition of chlorophenol and mineralization proceed almost simultaneously, and TOC (Total Organic Carbon) becomes zero.[56] With the treatment by ozone only, TOC does not decrease to nearly zero, but some amounts remain for a certain period. This indicates that chlorophenol decomposes to become different organic compounds but it does not decompose completely to CO_2 and intermediate compounds remain in the solution. The oxidative decomposition process of organic compounds with TiO_2 photocatalysts without ozone can be multistep reactions *via* several reaction intermediates, which may possibly produce toxic substances. By adding ozone, the concentration of such reaction intermediates can be substantially reduced.[56]

Furthermore, the cost of the electricity necessary for photoirradiation on TiO_2 photocatalysts and for the generation of ozone, together, is estimated to be only 5 cents per m^3 to treat the waste water containing TOC of $1\,mg\,L^{-1}$.[57] Hence, from the viewpoint of the cost, it is practical.

7.9 Effects of Surrounding Fields

7.9.1 Ultrasound

It is not possible to decompose water to oxygen and hydrogen by photocatalytic reaction with TiO_2 powder only. However, by applying ultrasonic wave (US) under argon atmosphere, water can be decomposed and hydrogen and oxygen are generated almost at the ratio of $2:1$. Because it is already known that only H_2O_2 and H_2 are generated on decomposing the water by ultrasonication, the reactions are considered to proceed as follows.

$$H_2O\ (US) \rightarrow {}^\bullet OH + H^\bullet \qquad (7.1)$$

$$2\ {}^\bullet OH \rightarrow H_2O_2 \qquad (7.2)$$

$$2\ H^\bullet \rightarrow H_2 \qquad (7.3)$$

$$2H_2O_2 + TiO_2(hv) \rightarrow O_2 + 2H_2O \qquad (7.4)$$

When the sonication and photoirradiation are alternatively performed, H_2 and H_2O_2 are initially generated by the sonication, and then

O_2 is generated by the following photoirradiation. In this case, in the presence of NaCl in the solution, water can be decomposed without a problem. Therefore, it can be applied to decomposition of sea water.[58] The activity of this reaction for rutile is higher than for anatase. And the activity becomes more enhanced for the relatively larger particle size. This would be attributed to the fact that the activity of photocatalytic decomposition of H_2O_2 for rutile is higher than for anatase and that the large particle size enhances the ultrasonic decomposition.

For the photocatalytic reaction by the combined use of TiO_2 and ultrasonication under CO_2–Ar atmosphere, the reduction of CO_2 proceeds and CO is generated.[59] Thus, for the photocatalytic system of the combined use of ultrasonication, the photocatalyst is used only for the oxygen generation as was seen in eqn (7.4). It is convenient for the separation and collection because H_2 and O_2 are generated separately, but because a large amount of electric energy is required for the generation of the ultrasonic wave, for practical applications, the energy efficiency in terms of the cost must be taken into account.

For the usual decomposition of the organic compounds, by combined use of ultrasonic decomposition and TiO_2 photocatalytic reaction, the synergistic effect that the reaction proceeds more rapidly than the summation of the individual reactions is observed. This would be attributed to the effect of H_2O_2 generated by the ultrasonic decomposition.[60]

7.9.2 Microwaves

The effect of microwaves on compounds and the applications to chemical reactions are also gathering attention. Because the energy of microwaves is too small to excite electrons across the bandgap of the semiconductor, they do not exert the redox reactions of photocatalysis. However, it is known that the activity is substantially enhanced.[61] With the apparatus shown in Figure 7.20A, decomposition of bisphenol-A (an environmental pollutant) is enhanced by microwaves as shown in Figure 7.20B,[62] where the reactant solution warmed by microwaves has to be cooled down to increase the efficiency.

As compared to that without microwaves, the reaction proceeds enormously rapidly, which would be attributed to the "hotspots" on the TiO_2 surface generated by microwaves. The remarkable increase in the decomposition rate in alcohols and in the suspensions of organic acids has been reported.[62]

For the photocatalytic reaction of ethylene in the gas phase, by combined use of microwaves, catalytic activity is sometimes

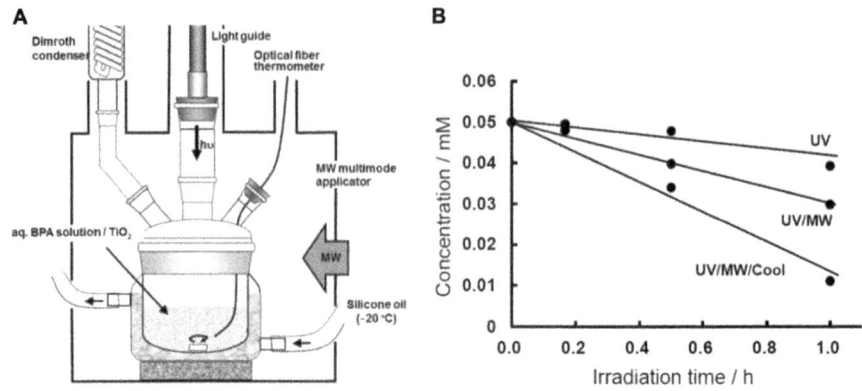

Figure 7.20   (A) Schematic illustration of the microwave-assisted photocatalytic reactor coupled to a cooling system. (B) Decrease
of the concentration of bisphenol-A in aqueous media by
photocatalytic oxidation (UV), by the microwave-assisted
photocatalytic oxidation (UV/MW), and by the microwave-
assisted photocatalytic degradation under cooling conditions
(UV/MW/Cool).
Reproduced from S. Horikoshi and N. Serpone, *Molecules*,
2014, **19**, 18102, MDPI.[61]

enhanced. This is related to humidity. Under low humidity, there was
no observed effect of microwave irradiation. By the application of
microwaves, the excessive adsorption of water on the TiO_2 surface
would be eliminated to enhance the photocatalytic activity.[63]

7.9.3   Magnetic Field

Because the photocatalytic reactions accompany the electron transfer
via radicals possessing electron spin, it is interesting to examine
whether the photocatalytic activity is enhanced by the effect of a
magnetic field. The effects of the magnetic field have been investigated since the initiation of photocatalytic research.[64] For the dye-
sensitized photocatalytic system, where ruthenium metal complex is
added to the system of TiO_2 deposited with Pt and RuO_2, on the
application of 0.4 T magnetic field under visible light irradiation,
the amount of hydrogen generation decreases by about 55%. On the
addition of methylviologen as an electron mediator, hydrogen
generation increases by 40 times without the application of the
magnetic field, while on the application of magnetic field, the
activity decreased by 55%. Based on these results, it is considered
that the magnetic field promotes the recombination reaction in
photocatalysts.[64]

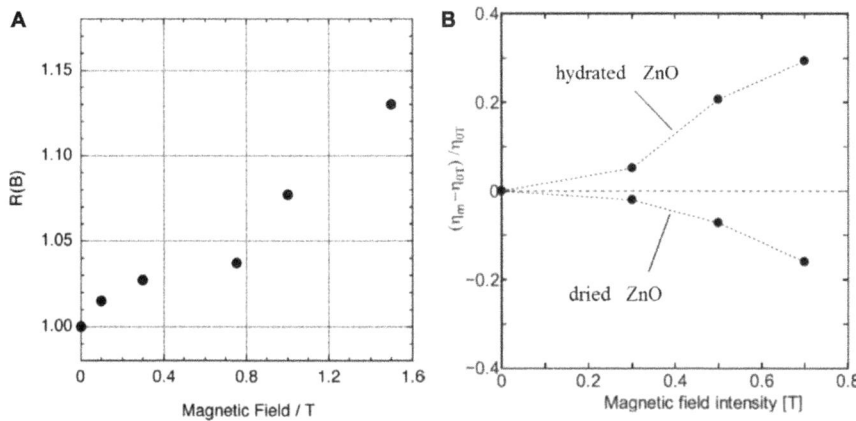

Figure 7.21 (A) Magnetic field dependence of relative reaction rate, *R(B)*, for the photocatalytic decomposition of *tert*-butyl alcohol with the ultrafine colloidal TiO$_2$. Reprinted with permission from ref. 65 (*J. Phys. Chem. B*). Copyright 2004 ACS. (B) Magnetic-field dependence on methylene blue photodegradation with two different ZnO particles. Reprinted from H. Okumura, S. Endo, S. Joonwichien, E. Yamasue and K. N. Ishihara, Magnetic field effect on heterogeneous photocatalysis, *Catal. Today*, **258**, 634–647. Copyright (2015), with permission from Elsevier.

Furthermore, as shown in Figure 7.21A, the decomposition rate of *tert*-butanol by TiO$_2$ increases by 13% on application of a magnetic field of 1.5 T.[65] For the decomposition of methylene blue in aqueous solution with TiO$_2$ photocatalyst, by taking enough time for the adsorption, the magnetic field does not affect the results. This could be explained by the fact that the magnetic field prevents the adsorption.[66]

For the nano-particulate ZnO possessing polarity as suggested in the crystal structure (see Figure 1.2), the dependency of the decomposition rate of methylene blue on the magnetic field intensity is shown in Figure 7.21B. The dependency is opposite for those with and without drying treatment. The reason has not yet been elucidated but it is reported that the effect of the paramagnetic oxygen molecules dissolved in the solution might not be neglected.[67]

References

1. Y. Nosaka, *J. Phys. Chem.*, 1991, **95**, 5054.
2. Y. Nosaka, N. Ohta and H. Miyama, *J. Phys. Chem.*, 1990, **94**, 3752.

3. H. Harada and T. Ueda, *Chem. Phys. Lett.*, 1984, **106**, 229.

4. Z. Li, R. Liu and Y. Xu, *J. Phys. Chem. C*, 2013, **117**, 24360.

5. L. X. Chen, T. Rajh *et al.*, *J. Phys. Chem. B*, 1997, **101**, 10688.

6. A. Nosaka, Y. Nosaka *et al.*, *J. Phys. Chem. B*, 2006, **110**, 8380.

7. Y. Nosaka, M. Nakamura and T. Hirakawa, *Phys. Chem. Chem. Phys.*, 2002, **4**, 1088.

8. H. Kominami and B. Ohtani, *Electrochemistry*, 1998, **66**, 996.

9. M. Cargnello, T. R. Gordon and C. B. Murray, *Chem. Rev.*, 2014, **114**, 9319.

10. P. Supphasrirongjaroen, P. Praserthdam *et al.*, *Ind. Eng. Chem. Res.*, 2008, **47**, 693.

11. M. C. Hidalgo, G. Cólon and J. A. Navio, *J. Photochem. Photobiol., A*, 2002, **148**, 341.

12. T. Wakamatsu *et al.*, *J. Jpn. Soc. Powder Powder Metall.*, 2001, **48**, 950.

13. H. Zhang and J. F. Banfield, *Chem. Rev.*, 2014, **114**, 9613.

14. B. Ohtani, Y. Ogawa and S. Nishimoto, *J. Phys. Chem. B*, 1997, **101**, 3746.

15. A. Sclafani and J. M. Herrmann, *J. Phys. Chem.*, 1996, **100**, 13655.

16. Y. Kakuma, A. Nosaka and Y. Nosaka, *Phys. Chem. Chem. Phys.*, 2015, **17**, 18691.

17. J. Zhang and Y. Nosaka, *J. Phys. Chem. C*, 2014, **118**, 10824.

18. R. Abe, H. Arakawa *et al.*, *Chem. Phys. Lett.*, 2001, **344**, 339.

19. M. A. Henderson, *Surf. Sci. Rep.*, 2011, **66**, 185.

20. J. Zhang, C. Li *et al.*, *Angew. Chem., Int. Ed.*, 2008, **47**, 1766.

21. R. Su, R. Bechstein *et al.*, *J. Phys. Chem. C*, 2011, **115**, 24287.

22. B. Ohtani, O. O. P.-Mahaney, D. Li and R. Abe, *J. Photochem. Photobiol., A*, 2010, **216**, 179.

23. D. C. Hurum, K. A. Gray, M. C. Thurnauer *et al.*, *J. Phys. Chem. B*, 2003, **107**, 4545.

24. D. O. Scanlon, J. Buckeridge *et al.*, *Nat. Mater.*, 2013, **12**, 798.

25. G. Liu, G. Q. Lu, H.-M. Cheng *et al.*, *Chem. Rev.*, 2014, **114**, 9559.

26. J. B. Lowekamp, G. S. Rohrer *et al.*, *J. Phys. Chem. B*, 1998, **102**, 7323.

27. T. Ohno, K. Sarukawa and M. Matsumura, *New J. Chem.*, 2002, **26**, 1167.

28. N. Roy, Y. Sohn and D. Pradhan, *ACS Nano*, 2013, **7**, 2532.

29. I. Paramasivam, P. Schmuki *et al.*, *Small*, 2012, **8**, 3073.

30. R. Asahi *et al.*, *Science*, 2001, **293**, 269.

31. K. Yamanaka, T. Ohwaki and T. Morikawa, *J. Phys. Chem. C*, 2013, **117**, 16448.

32. L. G. Devi and R. Kavitha, *RSC Adv.*, 2014, **4**, 28265.
33. S. Higashimoto *et al.*, *Appl. Catal., A*, 2008, **340**, 98.
34. R. Asahi *et al.*, *Chem. Rev.*, 2014, **114**, 9824.
35. J.-M. Herrmann, J. Disdier and P. Pichat, *Chem. Phys. Lett.*, 1984, **108**, 618.
36. U. G. Akpan and B. H. Hameed, *Appl. Catal., A*, 2010, **375**, 1.
37. H. Kato and A. Kudo, *J. Phys. Chem. B*, 2002, **106**, 5032.
38. M. Liu, X. Qiu, M. Miyauchi and K. Hashimoto, *J. Am. Chem. Soc.*, 2013, **135**, 10064.
39. H. Yu, H. Irie *et al.*, *J. Phys. Chem. C*, 2010, **114**, 16481.
40. M. Liu, M. Miyauchi *et al.*, *ACS Nano*, 2014, **8**, 7229.
41. S. Liu, M. Jaroniec *et al.*, *Adv. Colloid Interface Sci.*, 2012, **173**, 35.
42. C. Minero, E. Pelizzetti *et al.*, *Langmuir*, 2000, **16**, 2632.
43. Y. Xu *et al.*, *J. Phys. Chem. C*, 2007, **111**, 19024.
44. J. G. Yu *et al.*, *J. Phys. Chem. C*, 2009, **113**, 6743.
45. L. Korosi and I. Dekany, *Colloids Surf., A*, 2006, **280**, 146.
46. L. Jing *et al.*, *Energy Environ. Sci.*, 2012, **5**, 6552.
47. M. R. Gholipour, T.-O. Do *et al.*, *Nanoscale*, 2015, 7, 8187.
48. Z. Zhang and J. T. Yates, Jr., *Chem. Rev.*, 2012, **112**, 5520.
49. H. Saito and Y. Nosaka, *J. Phys. Chem. C*, 2014, **118**, 15656.
50. H. Cheng, H. Yamashita *et al.*, *J. Mater. Chem. A*, 2015, **3**, 5244.
51. Y. Nosaka, Y. Ishizuka and H. Miyama, *Ber. Bunsen-Ges. Phys. Chem.*, 1986, **90**, 1199.
52. Y. Nosaka, K. Norimatsu and H. Miyama, *Chem. Phys. Lett.*, 1984, **106**, 128.
53. H. Yoneyama and T. Torimoto, *Catal. Today*, 2000, **58**, 133.
54. T. Ohno, M. Matsumura *et al.*, *J. Catal.*, 2001, **204**, 163.
55. A. Mills, S.-K. Lee and A. Lepre, *J. Photochem. Photobiol., A*, 2003, **155**, 199.
56. P. Kopf, E. Gilbert and S. H. Eberle, *J. Photochem. Photobiol., A*, 2000, **136**, 163.
57. H. Noguchi and S. Sato, *Photocatalysis*, 2000, **1**, 47.
58. H. Harada, *Int. J. Hydrogen Energy*, 2001, **26**, 303.
59. H. Harada, C. Hosoki and M. Ishikane, *J. Photochem. Photobiol., A*, 2003, **160**, 11.
60. Y. Naruke, H. Tanaka and H. Harada, *Electrochemistry*, 2011, **79**, 826.
61. S. Horikoshi and N. Serpone, *Molecules*, 2014, **19**, 18102.
62. S. Horikoshi, M. Abe and N. Serpone, *Appl. Catal., B*, 2009, **89**, 284.

63. S. Kataoka, M. A. Anderson *et al.*, *J. Photochem. Photobiol., A*, 2002, **148**, 323.
64. J. Kiwi, *J. Phys. Chem.*, 1983, **87**, 2274.
65. M. Wakasa *et al.*, *J. Phys. Chem. B*, 2004, **108**, 11882.
66. S. Joonwichien *et al.*, *Appl. Magn. Reson.*, 2012, **42**, 17.
67. H. Okumura *et al.*, *Catal. Today*, 2015, **258**, 634.

8 Fabrication of Practical Photocatalysts

8.1 Preparation of Fine Particles

The crystallization of TiO_2 at low temperature is difficult, but at high temperatures sintering proceeds. Hence the fabrication of the TiO_2 nanoparticle is not straightforward. The processes to fabricate the powder nanoparticles are roughly categorized into three methods, *i.e.* solid phase method, solution phase method, and gas phase method. With the solid and solution methods, the problem is that contamination during the processes easily takes place. Furthermore, because with a solid phase method thermal hysteresis is large and the particles grow by calcinations, it is not suitable to obtain nanoparticles. Hence, the solution phase method is often adopted.[1]

With a solution method, because of the solvent, a liquid bridge is formed during the process in which the slurry is condensed and dried. As shown in Figure 8.1A, at the boundary between liquid (L), solid (S), and gas (V), the surface tensions γ_{LS}, γ_{LV}, and γ_{SV} work and the relationship described with the contact angle θ which is formed by the LV boundary and the solid surface holds. The penetration of the liquid among the spherical solid particles of radius a causes capillary condensation as shown in Figure 8.1B. Then, the capillary pressure p which causes the liquid bridge is given by the following equation,

$$p = \gamma_{LV}\left(\frac{1}{r_2} - \frac{1}{r_1}\right) \tag{8.1}$$

Introduction to Photocatalysis: From Basic Science to Applications
By Yoshio Nosaka and Atsuko Nosaka
© Yoshio Nosaka and Atsuko Nosaka, 2016
Published by the Royal Society of Chemistry, www.rsc.org

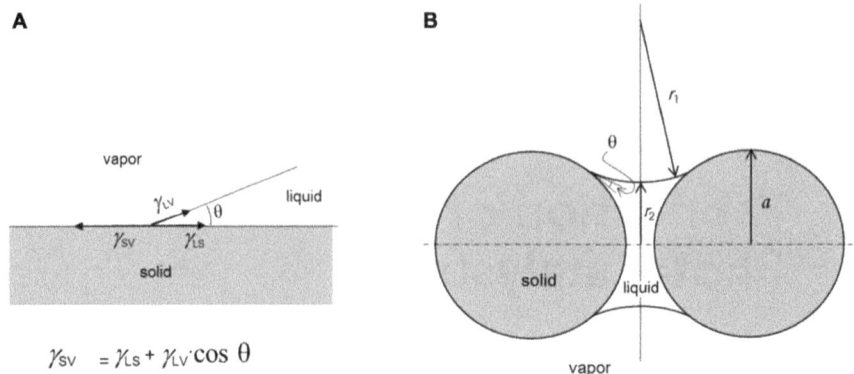

Figure 8.1 (A) Surface tensions γ_{SV}, γ_{LS}, and γ_{LV} among three phases (vapor, liquid, and solid) and the contact angle θ. (B) Liquid bridge between two particles of radius a to form capillary force.

where, r_1 and r_2 which determine the shape of the capillary bridge are given by the following differential equation,[2]

$$\frac{1}{r_2} - \frac{1}{r_1} = \frac{d(S_{LV} - S_{LS}\cos\theta)}{dV} \tag{8.2}$$

where V is the volume of liquid, and S_{LV} and S_{LS} are the areas of liquid–vapor and liquid–solid boundaries, respectively. Because $r_1 > r_2$ holds generally for the shape of the capillary bridge, the smaller the particle size is, p becomes larger. Namely, for nanoparticles, the degree of condensation becomes larger, and then the treatments become difficult. Accordingly nanoparticles with less condensation can be more easily obtained with a gas phase method which does not use solvent.

By use of a sol–gel method, nanoparticles can be readily fabricated at relatively low temperature. Historically, the sol–gel method was initially developed for TiO_2 nanoparticles to obtain high yields. The nanoparticles of TiO_2 can be prepared by washing the gel of the titanium alkoxides obtained after hydrolysis with water and they are then dried and calcined. The crystallinity increases after thermal treatments. Because of the poor quality of the as-prepared materials, hydrothermal and solvothermal methods were alternatively selected in place of the sol–gel method to prepare nanoparticles. All these procedures have been improving in recent years. Later, the development of surfactant-assisted non-hydrolytic routes has marked a remarkable improvement in this field to control the size and shape of TiO_2 for the preparation of tailored titania nanostructures.[3]

By use of the hydrolysis of metal alkoxides various metal oxides are fabricated such as TiO_2. With this alkoxide method TiO_2 film with high purity can be prepared, which will be described later.

The method by which the hydrolysis is carried out at high temperature and under high pressure is called a hydrothermal method. Similarly to the sol–gel methods, most of the hydrothermal approaches do not provide materials with tailored properties such as solubility, uniformity, and processability. In particular, in the absence of surfactants/protecting agents, the resulting particles are generally insoluble in common polar or non-polar solvents. Nevertheless, the crystallinity of the particles is usually much improved as compared to that by the sol–gel methods, and these materials can be very useful for some applications where dispersibility is not required. In addition, the hydrothermal methods require a few steps.[3]

A solvothermal method is very similar to the hydrothermal method except that the primary solvent used is not water although water is sometimes added to induce hydrolysis. In contrast to the hydrothermal method, a larger variety of surfactants or structure directing agents can be selected in the solvothermal method to drive the shape and morphology of the crystallites.[3] An example for the applications of the solvothermal methods is that the nanoparticles of anatase crystal phase could be obtained by the thermal treatment of titanium oxyacetylacetonate ($TiO(CH_3COCHCOCH_3)_2$) in glycol.

The above-mentioned methods would be sufficient to prepare photocatalytic nanoparticles at a laboratory level but it is not suitable for the practical preparation. Namely, to manufacture the photocatalyst products at industries, the process of the immobilization of the nanoparticles on the substrates is required. The preparation of photocatalytic powders for that process is difficult in terms of the cost.

8.2 Preparation of Thin Films by Wet Methods

8.2.1 Sol–Gel Method

TiO_2 thin films of crystalline anatase and rutile are prepared by using titania sol (the slurry of TiO_2 powder which is already crystalized) or non-crystal alkoxide precursor of TiO_2. For the former, solid particles with size of 1–100 nm are dispersed in the fluid medium, which is a kind of colloid. The gel is formed when the particles are assembled to form network structures and becomes solidified. For the latter case, which is called a sol–gel method, the film is prepared from liquid state, and a transparent, strong and adhesive film can be fabricated, though usually heat treatment is indispensable.

With a sol–gel method, the sol in which metal oxides and hydroxides are dispersed is prepared by hydrolysis and polymerization of the metallo-organic compound and the inorganic compound in the solution. Afterwards, by proceeding the reaction for gelation and heating, amorphous and porous crystals are fabricated. For the sol–gel method using alkoxides as raw materials, raw materials of high purity must be used and the product is fabricated at relatively low temperatures. During the fabrication process in the liquid states, different kinds of raw materials can be homogeneously mixed but pore, carbon, and OH groups remain. In addition, because inorganic films are fabricated from organic films, the substantial decrease in volume is a problem.

The metal alkoxide which is used as raw material is the compound in which metal binds with the oxygen of the alcohol. It is a kind of metallo–organic compound and is easy to treat differently from the highly reactive organo–metallic compound in which the carbon atom directly binds with metal. Because titanium alkoxide is reactive with water, hydrolysis reactions such as hydroxylation, dealcoholation, and dehydration, *i.e.*, eqn (8.3), (8.4), and (8.5), take place even in water under atmospheric conditions.[1]

$$Ti(OR)_4 + H_2O \rightarrow Ti(OR)_3 + ROH \tag{8.3}$$

$$Ti(OR)_4 + Ti(OR)_3OH \rightarrow Ti(OR)_3\text{-O-}Ti(OR)_3 + ROH \tag{8.4}$$

$$2\ Ti(OR)_3OH \rightarrow Ti(OR)_3\text{-O-}Ti(OR)_3 + H_2O \tag{8.5}$$

Depending on the amount of water, the dealcoholation, dehydration, and condensation reactions proceed at different ratios. Because these three reactions proceed in parallel, separation of the reactions is difficult. When the chemical structure of the alkyl group R of $Ti(OR)_4$ is different, the properties of titanium alkoxide and hydrolysis become different. When the number of carbon atoms of R is the same but the steric structure is different, the reactivity becomes different. Furthermore, when the structure is the same but the chain length of the R group is different, the reactivity also becomes different.

TiO$_2$ is fabricated by adding water of a 1.3 times larger amount than the alkoxide to the alkoxide. Then, among four R groups, three are rapidly hydrolyzed but the hydrolysis of the remaining group is slow. The reaction becomes slower with increasing the chain length, and the reactivity becomes lower with the increase of the polymerization. For stabilization, the preparation of acylate is carried out by reacting with carboxylic acid. On adding two times the amount of carboxylic acid, the following reaction proceeds.

$$Ti(OR)_4 + 2R'COOH \rightarrow Ti(OR)_2(OCOR')_2 + 2 \text{ ROH} \tag{8.6}$$

However, on adding three times the amount of carboxylic acid, the following complicated compounds are formed.

$$Ti(OR)_2(OCOR')_2 + R'COOH \rightarrow Ti(OR)(OCOR')_3 + ROH \tag{8.7}$$

$$Ti(OR)(OCOR')_3 \rightarrow O{=}Ti(OCOR')_3 + R'COOH \tag{8.8}$$

$$Ti(OR)(OCOR')_3 + Ti(OR)_2(OCOR')_2 \rightarrow (RO)_2TiOTi(OR)_2 + R'COOR \tag{8.9}$$

The stabilization methods suitable for the decomposition temperature must be selected on adopting chelate compounds as dipping solution such as β diketone, hydroxylcarboxylic acid, β ketoester, and glycol, which show different reactivity. Because Ti is a tetravalent and 6-coordinated transition metal possessing vacant d-orbitals, it readily forms a 5- or 6-membered chelate with the compound containing electron-donating substituents such as O, N, and S, with formation energy of 40–60 kJ mol^{-1}.

By hydrolyzing titanium alkoxides Ti(OR)$_4$, the sol of TiO$_2$ of particle size less than 1 nm can be prepared. On further processing of the hydrolysis, the primary particles are bound up to form a gel as depicted in Figure 8.2. When the sol with fluidity is coated on the heat-resistant substrates (glass, tile, or alumina) and calcined above 500 °C, gelation takes place to provide a transparent thin film, which is adhered on the substrates.

As coating methods to spread a precursor sol on the substrate, a dip coating method, a spin coating method, a roll coating method, and a screen printing method have been developed, as will be described below.

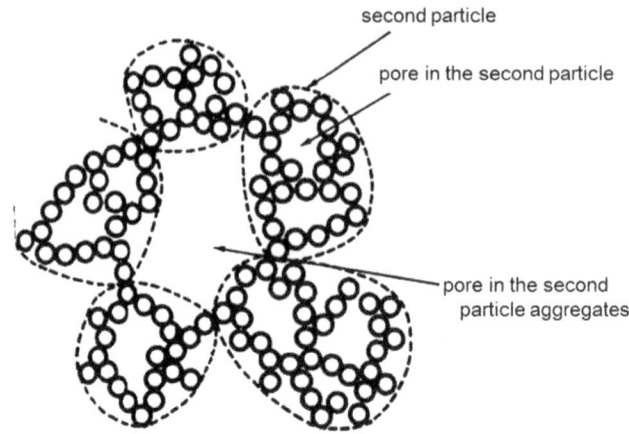

Figure 8.2   Gel formation by the aggregation of second particles.
Reproduced from ref. 4 (*Zoru-Geru Hou no Kagaku*, 1988) with
permission from Agune-Shofusha.

8.2.2   Dip Coating

The dip coating method is suitable for coating the substrate of a large
area. With this method a film of homogeneous thickness and superior outside appearance can be obtained in large quantities. To
fabricate the TiO_2 thin film with a dip coating method, titanium
isopropoxide is solved in alcohol and water is added to prepare the
sol. The substrate is dipped in the sol and dried. Afterwards it is
calcined and TiO_2 film is crystallized. Five steps are required from
dipping to drying as shown in Figure 8.3A.[5]

① Immersion; dipping the substrates in the sol.
② Pulling up the dipped substrates.
③ Deposition; the substrate is completely pulled up from the
   liquid to fix in the air.
④ Drainage; flowing down the excessive liquid on the substrates.
⑤ Evaporate the solvent and dry the substrate sufficiently. The
   hydrolysis proceeds simultaneously.

By pulling up the substrate at the constant rate, the sol can be
homogeneously coated on the substrate and the part above the liquid
surface is gelated to be immobilized. Thus, through evaporation and
gelation of the solvent on the surface of the substrate the condensation proceeds to form films. A continuous coat method shown in
Figure 8.3B is used for long vinyl sheets.

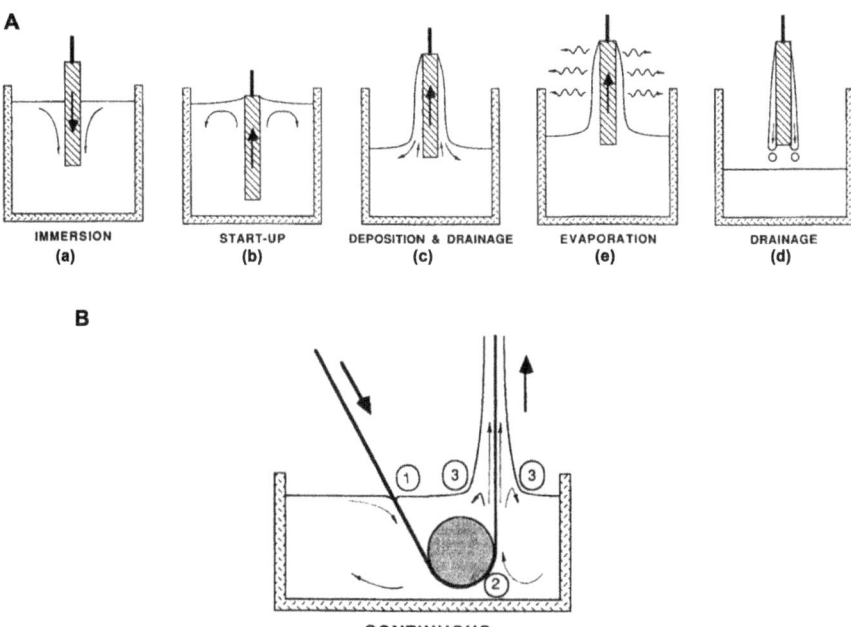

Figure 8.3   (A) Stages of the dip coating process: (A) batch and (B) continuous. Reprinted from C. J. Brinker and G. W. Scherer, *Sol–Gel Science*, ch. 13, pp. 786–837. Copyright (1990), with permission from Elsevier.

When the liquid viscosity (η) and substrate speed (u) are high enough to lower the curvature of the meniscus, the deposited film thickness (h) balances the viscous drag ($\propto \eta u/h$) and gravity force ($\rho g h$) as eqn (8.10),

$$h = c_1 \, (\eta u/\rho g)^{1/2} \tag{8.10}$$

where the proportionality constant c_1 is about 0.8 for Newtonian liquids.[5] When the substrate speed and liquid viscosity are not high enough, as is often the case in sol–gel processing, this balance is modulated by the ratio of viscous drag to liquid–vapor surface tension (γ_{LV}), and the relationship eqn (8.11) is suggested. By substituting eqn (8.11) for eqn (8.10), eqn (8.12) is obtained.[5]

$$c_1 = 0.94 \, (\eta u/\gamma_{LV})^{1/6} \tag{8.11}$$

$$h = 0.94 \, (\eta u)^{2/3}/\gamma_{LV}^{1/6}(\rho g)^{1/2} \tag{8.12}$$

Brinker and Ashley investigated the relationship between film thickness and withdrawal speed for a variety of silicate sols in which the precursor structures ranged from rather weakly branched "polymers" to highly condensed particles. Their results plotted in Figure 8.4 imply that, for polymeric systems, h varies approximately with $u^{2/3}$ in accordance with eqn (8.12).[5]

On adopting the sol–gel method using titanium isopropoxide, by selecting the suitable solution for dipping and calcining at 600–800 °C, after drying the films of anatase TiO_2 with almost constant refractive index (actually, the state in which anatase crystallites are dispersed in the amorphous) can be obtained. By adding additives, the film properties such as pore size and porosity can be controlled. However, the photocatalytic activity for decomposition alters depending on the amount of the additives.[1]

The coating reagents of low temperature cure, which do not require heat treatment, are also developed in order to produce the photo-catalyst films on the surface of the substrate of poor thermal resistance such as polyester films, general purpose resins, and paper products. For this case, firstly the substrate is immersed in the solution for the adsorption layer, then pulled up and dried for instance at around 70 °C, and is next immersed in the solution for photo-catalyst layer and pulled up and dried. When these two coating procedures are carried out, the films of high catalytic activity can be fabricated at relatively low temperatures.[1]

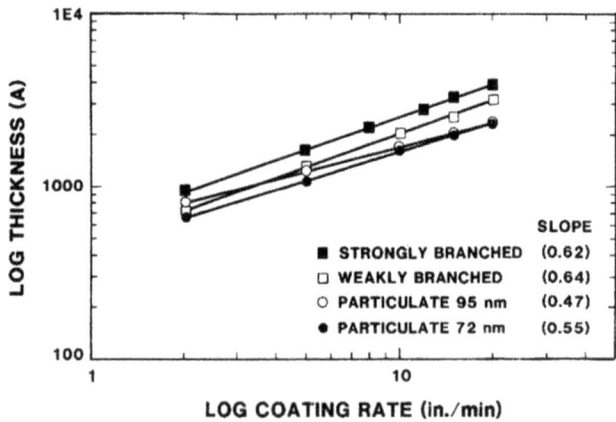

Figure 8.4   Coating thickness *vs.* coating rate for a variety of "polymeric" and particulate silicate sols plotted according to eqn (8.12). Reprinted from C. J. Brinker and G. W. Scherer, *Sol–Gel Science*, ch. 13, pp. 786–837. Copyright (1990), with permission from Elsevier.

8.2.3 Spin Coating

The spin coating method is effective for coating one side of small film materials. For TiO$_2$ films, a sol liquid of TiO$_2$ prepared by hydrolyzing titanium isopropoxide is condensed at room temperature with a rotary evaporator until it becomes viscous.[1]

Figure 8.5 shows the procedures of spin coating, which consist of four steps, that is ① deposition, ② spin-up, ③ spin-off, ④ evaporation.[5] Details are as follows:

① After keeping the substrate on the device (spin coater) which can rotate at 1000–3000 r.p.m. an excessive amount of sol liquid is dripped on the center of the spin coater so that the whole substrate can be covered.

② Afterwards, the spin coater is rotated at around several hundreds of r.p.m. When the liquid homogeneously covers the substrate, the rotating is switched to a faster speed.

③ At around several thousand r.p.m., the sol liquid rapidly spreads by centrifugal force and the substrate is homogeneously coated.

④ Then, the solvent is evaporated and successively gelation proceeds.

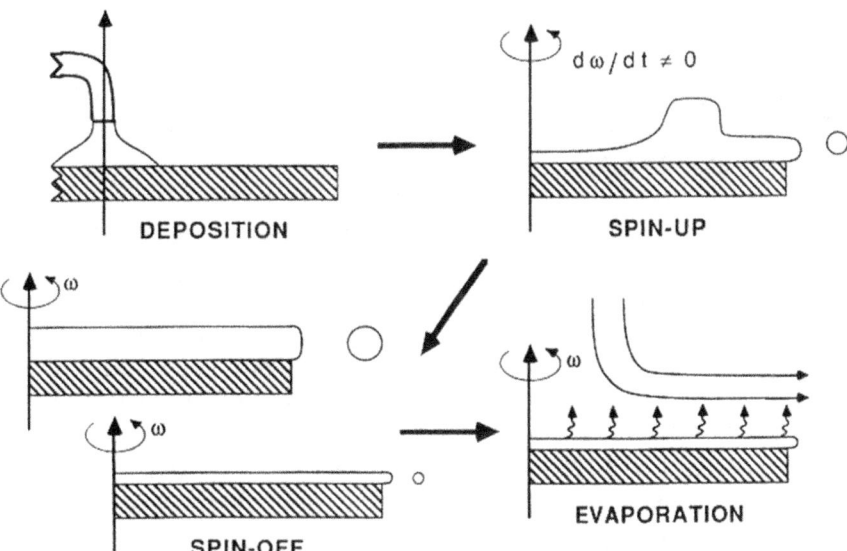

Figure 8.5 Stages of the spin coating process.
Reprinted from C. J. Brinker and G. W. Scherer, *Sol–Gel Science*, ch. 13, pp. 786–837. Copyright (1990), with permission from Elsevier.

When the substrate with the initial homogeneous film thickness of h_0 is rotated at the angular velocity ω for t seconds at the spin-off step ③, the thickness of the film $h(t)$ can be expressed by the following equation,[5]

$$h(t) = h_0/(1 + 4\rho\omega^2 h_0^2 t/3\eta)^{1/2} \qquad (8.13)$$

where η and ρ are the viscosity and the density of the solution, respectively. Even films that are not initially uniform, sooner or later monotonically tend toward uniformity following eqn (8.13).[5] As indicated in this equation, the film thickness is determined depending on the viscosity of the solution, the aging period until the actual usage after the preparation, and rotation velocity, *etc.* Transparent and porous TiO_2 films are fabricated by the heat treatment of the coated gel films at around 400–500 °C. The heat treatment temperature is an important factor to determine the crystallinity and photocatalytic activity of TiO_2. When the film by one heat treatment process is thick, cracking in the film and separation from the substrate can take place. Therefore, to control the film thickness, the spin coating and heat treatment processes are alternately repeated.

8.2.4 Insufflation Method

The insufflation method is a method to manufacture films by spraying the sol liquid in a mist or vapor state, which are prepared with titanium organic compounds such as titanium alkoxide or titanium chelates on the substrate heated above 400 °C. A pyrosol process and a spray process are known.

A pyrosol process is adequate for the fabrication of oxide films for the multi-component system which contains various kinds of components in the solution. As shown in Figure 8.6, by placing the precursor solution of TiO_2 in the ultrasonic humidifier and generating an ultrasonic wave, a mist is formed. The mist is transferred onto the substrate heated with carrier gas and subjected to pyrolysis. Owing to the ultrasonic wave, the distribution of the particle size of the mist becomes extremely narrow. When the frequency of the ultrasonic wave is 3 MHz, the particle size of mists becomes 2 μm. The film quality depends on the heating temperature of the substrate.[1] By dispersing ultrasonically a suspension of P25 TiO_2 in water, the film can be also fabricated on the glass plate by the pyrosol process.

In a spray process, titanium alkoxide and water are sprayed on the substrate heated at a certain temperature to obtain TiO_2 thin films.

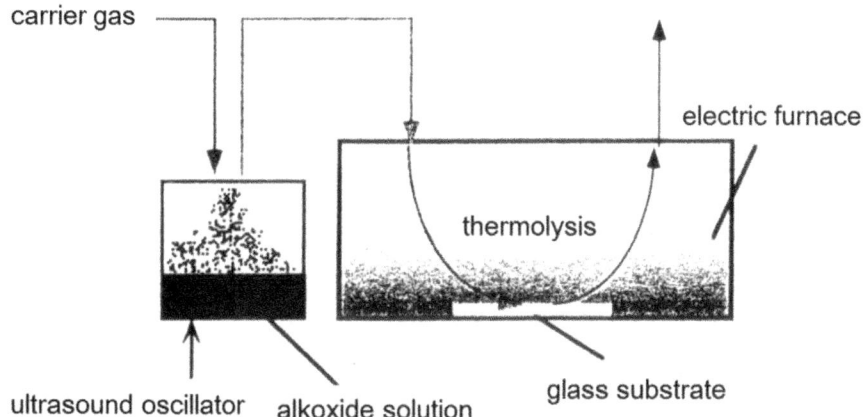

Figure 8.6 Pyrosol process for film fabrication.
Reproduced from ref. 1 (*Sankachitan Hikarishokubai no Subete*, 1998) with permission from CMC Publishing.

Because the substrate of curved shape can also be coated with this process, the process is adequate for providing glass products and lighting equipment with antibacterial and antifouling effects. By adopting $TiCl_4$ as starting material, thin films can also be fabricated. In principle this process is the same as the atmospheric pressure chemical vapor deposition, which is described later.

8.3 Preparation of Thin Film by Dry Methods

Dry methods have recently been applied industrially more frequently than the wet method. Spattering, CVD (Chemical Vapor Deposition), and thermal spraying procedures are popular.

8.3.1 Spattering

Spattering is the phenomenon whereby the atoms and molecules that form solids are emitted from the solid when the particles possessing enough energy collide with the solid surface in a vacuum. The instrument shown in Figure 8.7, with which the magnetic field is applied (or magnetron-spattering), is widely used for the fabrication of thin films.

For the fabrication of TiO_2 thin films in the vacuum apparatus, by using metal Ti as a target and Ar as spattering gas, the voltage is applied under oxygen atmosphere to collide the particles at high speed. Then, small pieces of titanium atoms are spattered, which are

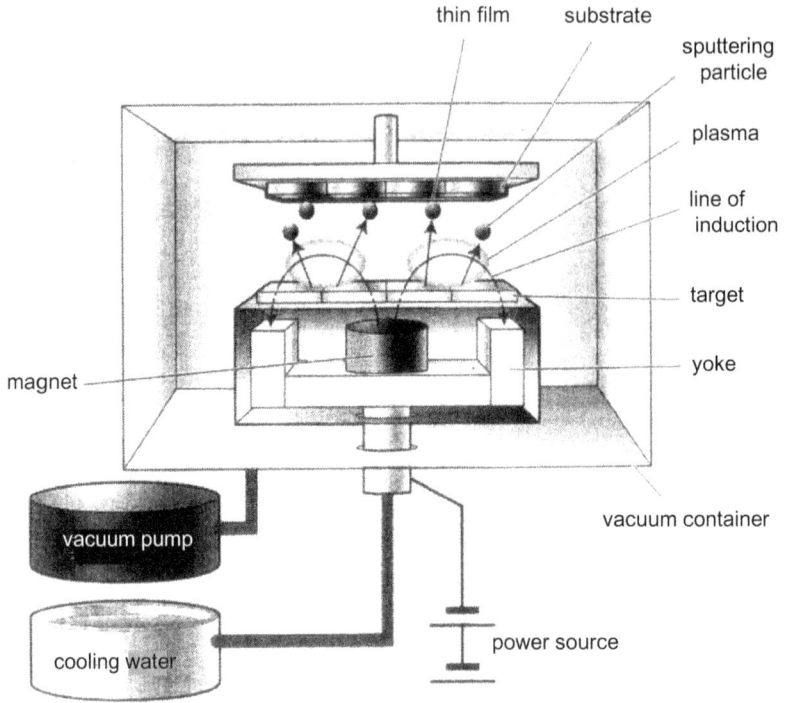

Figure 8.7   Principle of a magnetron-spattering method.
             Reproduced from ref. 6 (*Zukai Hikarishokubai no Subete*, 2012)
             with permission from Ohmsha Ltd.

transferred onto the substrate, and TiO_2 films are fabricated. By controlling the conditions on fabricating films, growth of TiO_2 films of rutile crystal form on the unheated substrate also becomes possible,[6] but to make the TiO_2 film with high photocatalytic activity the pressure of the gas is set higher in the atmosphere containing a large amount of oxygen. Furthermore, it is necessary to raise the temperature of the substrate, so that the crystallization of the TiO_2 becomes easier. This method has an advantage that the condense thin films can be homogeneously coated on the substrate of large area and is practically applied to coat the side mirrors of cars with TiO_2.

With the usual spattering procedure, the rate of film formation is about $1–2\ nm\ min^{-1}$. It is reported that by using a dual cathode spattering process controlled by plasma emission monitoring, and a direct current spattering procedure where $TiO_{1.986}$ (prepared by reducing TiO_2) is adopted as a target, the film formation rate becomes higher by more than one order of magnitude. The polycrystalline film which is fabricated at high formation rate and afterwards calcined at

200–400 °C reportedly presents sufficient photocatalytic activity and superhydrophilicity.[6] Furthermore, to fabricate films on the non-heat-resistant resin substrate an unbalanced magnetron spattering procedure (UBM) is reported to be effective.

8.3.2 Chemical Vapor Deposition

Chemical vapor deposition (CVD) involves either (i) chemical reaction of a volatile compound to be deposited, with other gases, to produce a non-volatile solid or (ii) pyrolysis (thermal decomposition) of a compound at high temperature to produce a solid. In either case the solid forms as a film on a suitable substrate. It is important that the reaction proceeds in the gas stream over the substrate, then particulates form. Under these conditions the film density becomes low and the film possesses a large number of voids. There are several forms of CVD and each form has its own acronym.[7] If the process takes place at atmospheric pressure (AP) it is referred to as APCVD. In APCVD systems the gas flow is almost exclusively parallel to the surface. When lower pressure is used the operation is low-pressure CVD (LPCVD). The gas pressure is usually in the range 0.5–1 Torr for LPCVD reactors while it is 760 Torr for an APCVD system, by which the CVD is distinguished. When plasma is used to generate ions or radicals that recombine to give the desired film, the process is plasma enhanced CVD (PECVD). For PECVD it is possible to use much lower substrate temperatures because the plasma provides energy for the reaction to proceed. Many nitrides have been prepared in thin-film form by PECVD, including AlN, GaN, TiN, and BN. For the preparation of TiO_2, $TiCl_4$ and O_2 gas are adopted as the reactants at substrate temperatures of 200–400 °C, where the deposition rate is about 0.1 nm s^{-1}.[7]

8.3.3 Thermal Spraying

The thermal spraying method has been developed as a surface treatment technique which is represented by a plasma thermal spraying process. The sprayed film is fabricated by depositing nanoparticles heated to high temperature on the surface of the substrate. This technique is characteristic because a thick film can be fabricated in a relatively short time. Because the particle size of the ceramic powder used for spraying is generally $5 \sim 50$ μm, in the case of TiO_2 the aggregated particles of size of about 40 μm which are prepared with a spray dryer are sprayed with high speed gas frame above 900 °C.

Therefore, part of the anatase crystals are transferred to rutile but the formed film presents the photocatalytic activity.[8]

8.4 Modification of Photocatalyst Particles

When TiO_2 powder is dispersed to be deposited on non-heat-resistant materials such as paper, cloth, fluororesin, and coating paint, TiO_2 must be exposed on the surface of the materials because the photocatalytic reaction is a surface reaction. It is also necessary to protect the materials from decomposition by the oxidative exertion of TiO_2.

Various modifications of TiO_2 particles have been proposed. As indicated in Figure 8.8, by a gas phase method a composite oxide where SiO_2 is placed on the surface of the primary photocatalyst particles without agglomeration can be fabricated by one step. For the procedure by which SiO_2 is partially deposited, $TiCl_4$ and $SiCl_4$ are used together, which are preheated at around 1000 °C under flowing oxidative gas. In this case because the reaction of the oxidative gas with $TiCl_4$ is more rapid than that with $SiCl_4$, TiO_2 particles are initially formed, then SiO_2 are formed. Therefore, most of the SiO_2 grows on the TiO_2 surface. By controlling the reaction conditions, the primary particle size and the content of SiO_2 can be controlled extensively.

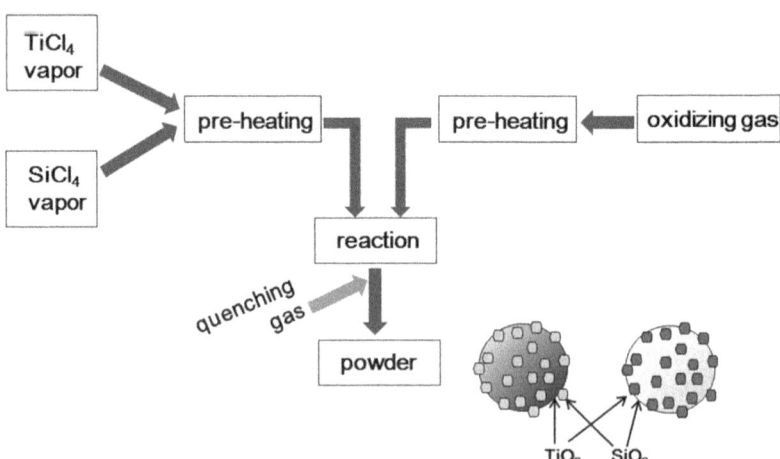

Figure 8.8 Manufacturing course of mixed oxide by gaseous method. Reproduced from ref. 9 (*Kaiho Hikarishokubai*, 2004) with permission from Photo Functionalized Materials Society.

When photocatalytic function has priority, the particle size and the amount of SiO_2 should be set small. On the other hand, when the weather resistance is emphasized, to the contrary the particle size should be set large and the content of SiO_2 should be increased. With such partial surface treatments one can fabricate well-balanced composite oxides with photocatalytic function and weather resistance corresponding to the purpose of the use. Thus, composite oxides have been developed as interlaced materials for resins and are practically utilized by depositing in fibers.[9]

Hydroxyapatite $(Ca_{10}(PO_4)_6(OH)_2)$ is a material which is contained in bones and teeth. It presents a superior adoption capability against biological substances such as nucleic acids and proteins. A composite nanoparticle photocatalyst where the surface of TiO_2 is covered with apatite like confetti, which is a star-shaped small rock candy, is also developed. On depositing dispersively on fibers, the apatite on the surface prohibits the contact of the fibers with TiO_2 to prevent decomposition of the fibers by TiO_2.

8.5 Development of Interlayer

When soda-lime glass is adopted as a substrate, TiO_2 films are fabricated at high temperature. Therefore, as shown in Figure 8.9, alkaline component is dispersed into the photocatalytic layers during the heat treatment process to decline the photocatalytic activity. Sodium in the soda-lime glass is dispersed to react with TiO_2 on the surface to form sodium titanate $Na_xTi_yO_z$. Hence, the effective amount of TiO_2 becomes less or the recombination center increases to decline the photocatalytic activity. To prevent such dispersion of sodium, SiO_2 is undercoated as an intermediate layer in advance. Without the undercoat of SiO_2, sodium disperses to a large extent

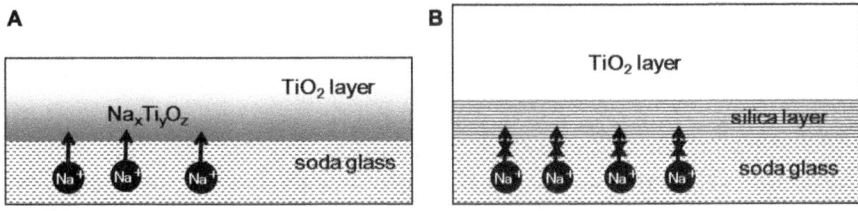

Figure 8.9 Diffusion of Na^+ to TiO_2 layer (A) is protected by silica protective layer (B).
Reproduced from ref. 1 (*Sankachitan Hikarishokubai no Subete*, 1998) with permission from CMC Publishing.

during the repeated processes to fabricate a thick film, and the transmittivity of visible light decreases. However, with undercoating, film of thickness more than 1 μm and transmittivity of more than 80% is attained on average. Thus, by introducing the intermediate layer, the fabrication of the transmissive films with high photo-catalytic activity can be achieved.[1]

On fabricating films of photocatalytic materials directly on the substrate by the above-mentioned dry or wet processes, the substrate must be heat resistant against high temperatures. However, as stated above, by adopting TiO_2 prepared beforehand and a suitable binder, films can be fabricated on the surface of various materials at relatively low temperatures. On immobilizing TiO_2 on the substrate, three layers (a binder layer, an interlayer, and a protecting layer) are introduced in order to make the film adhesion higher, to prevent the dispersion of impurities from the substrate and the photocatalytic deterioration of the substrate of organic compounds. The interlayer often provides the function to adhere the photocatalyst layer with the substrate, in addition to protect the substrate from deterioration by photocatalytic reactions. Some substrates introduce more than two layers of adhesion layers and interlayers to improve the durability and photocatalytic activity.

As a durable binder which is not oxidized by the strong oxidizing power of photocatalysts, a gel of extremely fine particles of SiO_2 is generally utilized. When the sol of silica nano-particles and fine particles of TiO_2 are mixed in the solution and painted on the surface of the substance, during solvent evaporation in the drying process, Si nanoparticles adhere to each other to become gel to form a binder, resulting in the generation of hard and strong thin films. With a sol–gel of silica, a strong film can be fabricated within the temperature range from room temperature to 150 °C.[1]

Coating reagents which contain photocatalytic fine particles to deposit photocatalysts on the surface of the substrate have been developed. In most cases, a double coat process is adopted by using two kinds of coat reagents, for the undercoat and final coat processes. The first layer composes an interlayer. As shown in Figure 8.10, when the substrate is an organic compound, the reagents to form organic and inorganic component-gradient film are selected. For instance, as an organic segment, copolymer of methyl–methacrylate and methacryl-oxypropyl-trimethyl-silane was considered, while as an inorganic segment partially hydrolyzed compound of tetra-ethoxy-silane was considered. When this organic–inorganic mixed polymer is coated and dried, the organic segment is adsorbed on the organic substrate, while

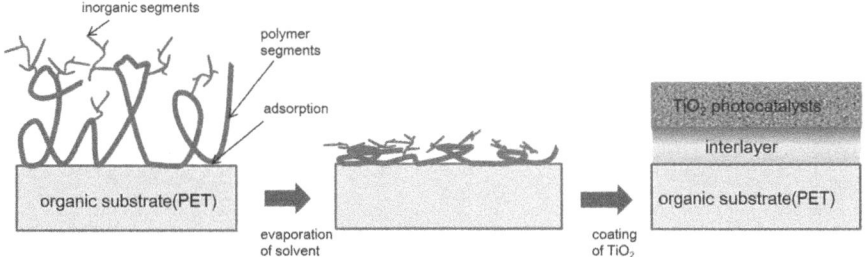

Figure 8.10 Two-coat photocatalytic film with an inorganic–organic gradient interlayer.
Reproduced from ref. 10 (*Hikarishokubai*, 2005) with permission from NTS Inc.

to the contrary, the inorganic segment appears in the outer layer and binds with the photocatalytic film in the second layer through binding with silica.[10]

Because the double layer coat takes time to process, it is contrived that the photocatalyst layer is separated on the surface by a single coating. For a single layer coat, as shown in Figure 8.11, two types of coating are proposed, (a) the photocatalyst particles gather around the surface and (b) the particles float from the surface.[11] On evaporating the solvent during the drying process, the thickness of the coat becomes one-tenth. It is processed so that along with the evaporation, the solvent makes the photocatalytic particles move to gather outside. For the float type, it is processed so that the photocatalytic particles float to the surface by the difference of surface tension.

8.6 Practical Applications to Representative Materials

8.6.1 Ceramic Tiles

Tiles are a very initial practical example to which TiO_2 photocatalysts were applied, which are produced according to the procedures shown in Figure 8.12.

TiO_2 is painted so that transparent and homogeneous films can be formed on the tiles. The firm films are formed with a baking technique with which TiO_2 fine particles can be partially buried in the glaze. Because the phase transition from anatase to rutile takes place on heating, the photocatalytic activity declines. To attain high photocatalytic activity the baking temperature should be set not to exceed the phase transition temperature.

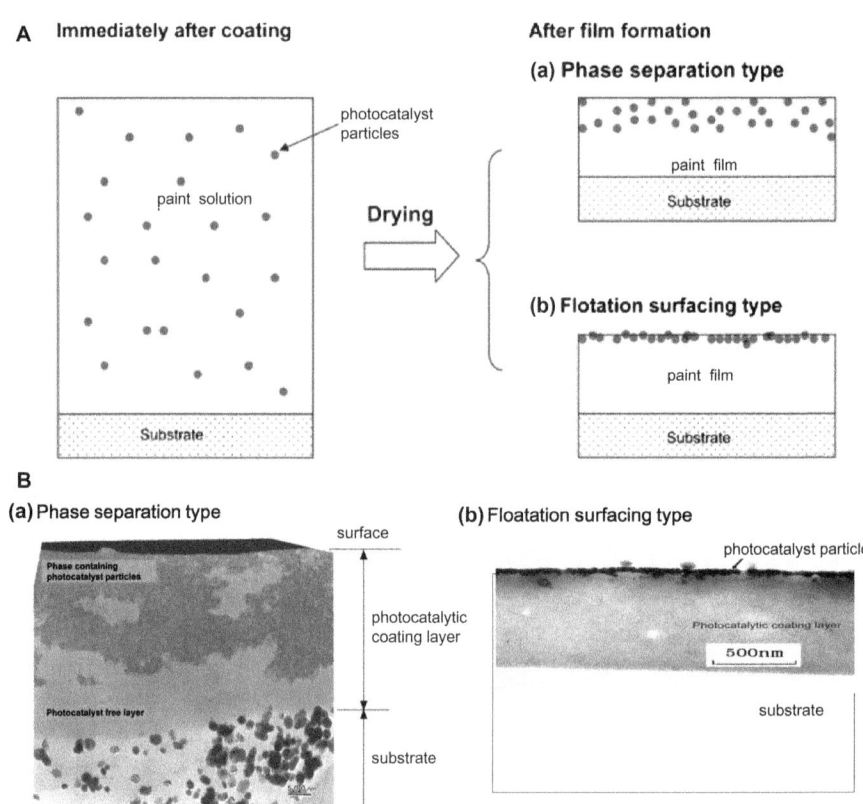

Figure 8.11 (A) Preparation process for (a) phase separation and (b) floating types of one-coat photocatalytic film and (B) the cross-sectional views for each type by electron microscopy.
Reprinted with permission from J. Oguma, T. Niguma, A. Nakabayashi and Y. Nosaka, Top-level performance. Particle flotation enhances efficiency of self-cleaning coating, *Eur. Coat. J.*, **12**, 84–87. Copyright (2011) Vincentz Network.

When solutions containing Ag or Cu ions are coated on the tiles of TiO_2 photocatalyst and are illuminated with ultraviolet light, Cu and Ag are immobilized on the surface of TiO_2 by the reductive effect of the photocatalyst TiO_2. Thus, a higher antibacterial effect is presented. Usually with the use of TiO_2 solely, an antibacterial effect is not exerted unless ultraviolet light with intensity of more than 10 $\mu W\,cm^{-2}$ is applied. However, with the combination with antibacterial metals, the photocatalytic activity and high antibacterial effect are realized under the illumination of light of 1 $\mu W\,cm^{-2}$.[12]

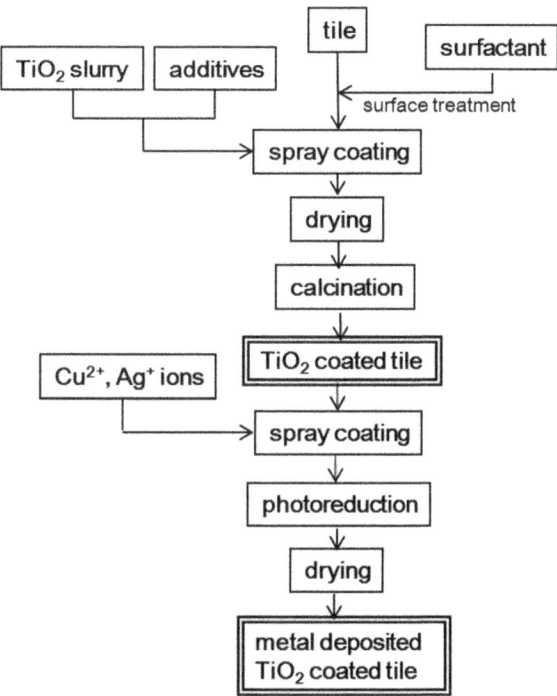

Figure 8.12 Manufacturing course of metal deposited TiO₂ photocatalyst coated tile.
Reproduced from ref. 12 (*Sankachitan Hikarishokubai*, 1998) with permission from CMC Publishing.

8.6.2 Glass

When the solution in which TiO₂ photocatalyst particles are dispersed in the inorganic substance whose main component is silica (SiO₂) is painted on the glass substrate very thinly and homogeneously, then calcined, a transparent film with excellent durability is fabricated. For mass production, thin films are efficiently fabricated in the process of sheet glass production (see Figure 8.13), by placing a CVD coater in the glass ribbon formation part with a float bath of molten tin.[6]

Because the refractive index is close to that of glass and the films are colorless and transparent, the appearances cannot be distinguished from the usual glass. However, on ultraviolet light illumination, the films exhibit superhydrophilicity and the ability to decompose organic compounds. Besides the self-cleaning function, TiO₂ coating on a window glass can be utilized to reduce the cooling power of the building by utilizing heat of evaporation by flowing a small amount of water on the surface as will be described in Chapter 10 (Section 10.6.2).

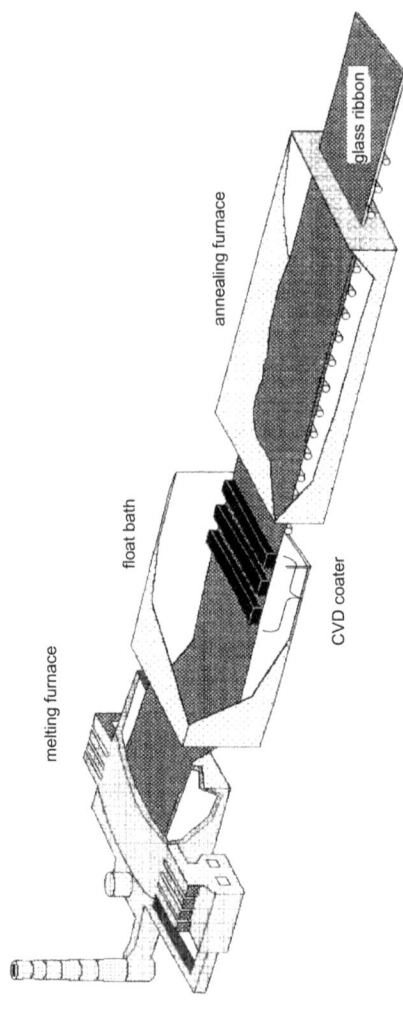

Figure 8.13 In-line CVD process for coating TiO$_2$ film on a sheet glass. Reproduced from ref. 13 (*Kinou Zairyou*, 2003) with permission from CMC Publishing.

8.6.3 Tents

For a middle-scale fabric building such as a tennis court, relatively inexpensive polyester or polyvinyl chloride (PVC) film is used with the fiber fabric. PVC film has the advantages that manipulations are feasible and various colors can be applied. However, it has disadvantages that dirt adheres easily because PVC film is itself lipophilic, and that the durable life is short. Therefore, to make dirt inconspicuous, it is often deeply colored. When the PVC film surface is coated with photocatalysts (Figure 8.14A), the stain adsorbed on the surface can be photocatalytically decomposed to be washed by rain. Thus, the antifouling property is excellent, and it is not necessary to color the tent sheet to conceal the fouling. It also has the advantage that the difference between outdoor and indoor temperatures does not become large because in daytime the intensity of illumination is almost satisfied with only the sunlight intensity. Thus, by adopting TiO_2-PVC films the antifouling effect can be improved and the cost required for air conditioning and lighting can be also reduced.[13]

After TiO_2-PVC film, a TiO_2-PTFE (poly-tetrafluoroethylene) film was developed, where the photocatalytic function is provided by mixing TiO_2 photocatalyst with fluororesin as shown in Figure 8.14B. The film materials in which glass fiber fabric is coated with PTFE film are used as fabric material. The PTFE film is a superior film material because it is stain resistant, the durable life is long (about 30 years), and it is certified as a non-combustible material. It is mainly applied for the dome roofs of large-scale fabric buildings. This film can deal with soot and smoke, which is not attained with PTFE film.

The decomposition ability of TiO_2-PTFE film is about twice as high as that of TiO_2-PVC film. However, TiO_2-PVC film presents hydrophilicity while TiO_2-PEFT film does not, but it keeps its water

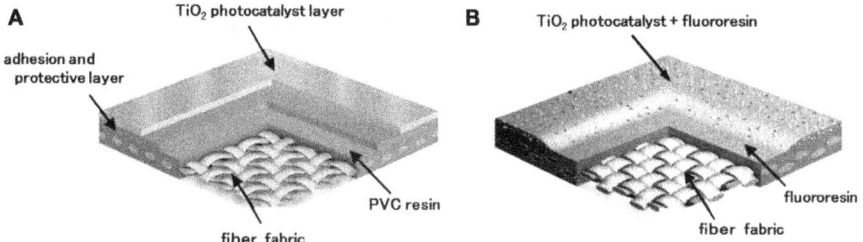

Figure 8.14 Two types of photocatalytic tent with (A) PVC resin and (B) fluororesin.
Reproduced from ref. 14 (*Zukai Hikarishokubai no Subete*, 2012) with permission from Ohmsha Ltd.

repellent property. Because the thin layer of water can be formed in the large area on the surface of PVC-film tent which presents hydrophilicity, the excellent evaporating cooling effect can be expected with a small amount of water. By exploiting this characteristic, the development of an outdoor cooling system with the photocatalyst-hydrophilic tent is also attempted.

8.6.4 Aluminum Building Materials

Most aluminum building materials are covered with an organic coating by methods such as electrodeposition of acrylic resin. Therefore, on forming photocatalyst films directly on the materials, due to the photocatalytic affects, the ground coat is decomposed and deteriorated. Hence, it is difficult to keep the durability. Because outdoor building materials are exposed to solar radiation, wind, rain, and snow, performance to bear the rigorous natural conditions is required for aluminum building materials. To fulfill the requirements, the adherence is improved by placing an interlayer between the organic coating and TiO_2 layer as shown in Figure 8.15. And the building material which possesses the antifouling ability of a photocatalyst and the durability of outdoor materials is developed. This aluminum building material is widely applied practically and

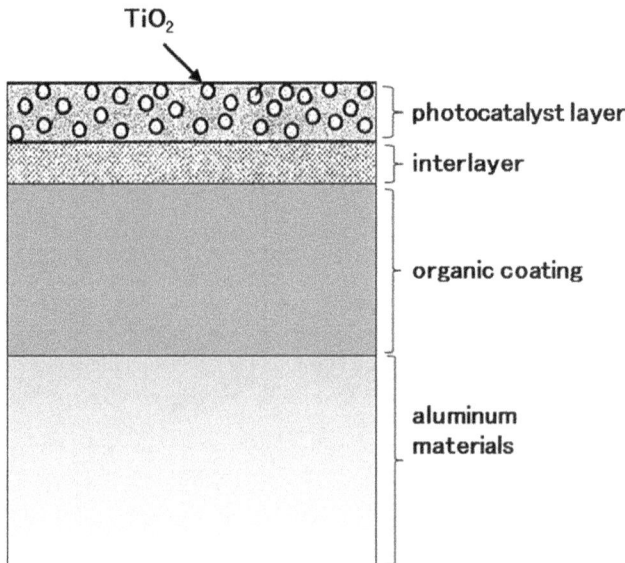

Figure 8.15 Photocatalyst-coated aluminum building materials.

recognized as useful to keep the beautiful appearance of the building and reduce the maintenance cost to clean buildings.[15]

8.6.5 Polymer Films

More than two interlayers are adopted to immobilize TiO_2 on the transparent polymer film substrates such as polyester and acrylic resin films to improve the adhesion between the substrate and TiO_2 and to prevent the deterioration of the substrate. In the case of polyester film, firstly the adhesive layer is coated on the substrate of polyester film to make the adhesion of the inorganic coating feasible, then by coating the inorganic interlayer the film is made harder. Furthermore, inorganic compounds, TiO_2, and silica layers are additionally coated. The product of a multi-layer structure was manufactured previously, but recently by adopting a single coating process the coating process has become simple and the transparency and the durability are improved as described in Section 8.5. The film fabricated in this way is used for window film by coating adhesive agents on the back.[16]

The materials in which TiO_2 is coated on the surface of the plastic film are practically applied for the films to keep their freshness. Because by use of this film the maturation hormone, ethylene gas, which is emitted from fruits and vegetables can be decomposed to carbon dioxide and water by the photocatalytic exertion, the maturity of foods can be delayed, and rot can be prevented to keep the freshness.

8.6.6 Cloths and Papers

When TiO_2 photocatalyst is directly incorporated in fibers, due to the photocatalytic effect the fibers are decomposed. Therefore, the application is difficult. To circumvent the problem, TiO_2-hybrid photocatalysts deposited with ceramics such as apatite, which is photocatalytically inactive, were developed. Though the photocatalyst does not present the functions unless light is irradiated, the TiO_2 photocatalysts with apatite adsorb bacteria and mold to prevent the growth without light. The composite TiO_2 fine particle photocatalysts of confetti shape, which was described in Figure 8.8, are practically applied to clothes, curtains, carpets, sportswear, hair brushes, masks, bedclothes, handrails, *etc.*

The pulp which comprises paper is a superior dispersive carrier of TiO_2. As shown in Figure 8.16, on adding flocculating materials in the TiO_2 sol, TiO_2 is flocculated, which is mixed with pulp. Then the

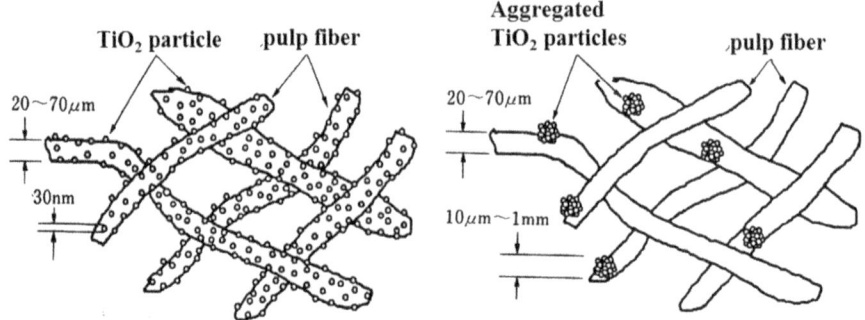

Figure 8.16 Schematic illustration for two types of deposition of photo-catalyst powders on papers.
Reprinted from ref. 17 with permission from Industrial Technology Center, Gifu.

flocculated TiO_2 (about several tens of μm in size) is deposited on the pulp as shown in Figure 8.16. Because the contact area of TiO_2 with the pulp is reduced on the flocculation of TiO_2, the strength of the whole paper is retained.[17] Because pulp is an organic compound, the pulp around the flocculated TiO_2 seems to be decomposed by photocatalytic exertion, leading to the decomposition of fibers and the unloading of powders. However, because for the actual paper the fibers of the pulp are piled up by several thousands or several tens of thousands of layers, the sparsely distributed TiO_2 aggregates do not degrade the strength of the paper.

On the other hand, TiO_2-hybrid photocatalyst particles in which photocatalytically inert ceramics are attached on the surface of TiO_2 are coated or incorporated for papers so that decomposition of papers is prevented because TiO_2 does not have direct contact with the paper. These papers are manufactured and practically utilized for sliding screen (shoji) papers, wallpapers, and lamp shades, *etc.*

8.6.7 Air Cleaner

The TiO_2 photocatalytic materials with the three-dimensional mesh ceramic form as shown in the insert photo in Figure 8.17 are developed and mounted on the environment cleaning apparatus. This three-dimensional mesh ceramic form possesses the following characteristics:

① Because the area of the framework is large, a large amount of TiO_2 can be deposited.

② The generated sulfates and nitrates can be catalyst poison when inorganic compounds such as H_2S, SO_X, NH_3, and NO_X are decomposed with photocatalysts, but these can be removed by washing. The highly concentrated organic compounds adsorbed on the surface can also be removed by heat treatment.

③ The distance from the light source can be shorter because it is heat resistant.

④ Because the porosity is very high (80–90%), the ventilation resistance is extremely small.

⑤ Because the framework is random, the contact efficiency with fluids is good.

Because the shape of the filter is unique, even without photo-irradiation 30–40% of saprophytic bacteria can be captured. Fifteen minutes after the photoirradiation, the survival rate of the floating fungi decreases down to 17%.[18]

Recently, a lot of air cleaners equipped with a photocatalyst filter have been manufactured. An example of the commercial air cleaners for residential use is shown in Figure 8.17. For this air cleaner, dust and rubbish are firstly removed by the pre-filter on the front. The particulate substances such as house dust and cigarette smoke which can pass the filter are removed by the next HEPA filter. Successively,

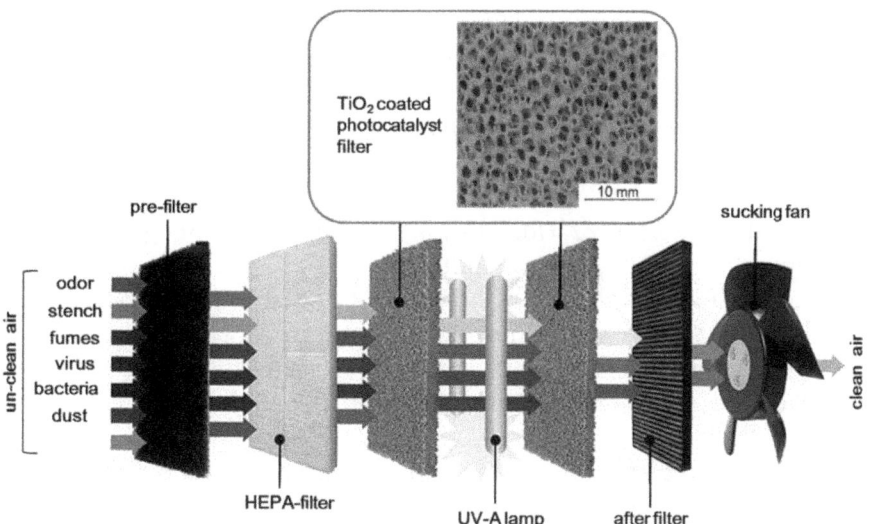

Figure 8.17 Photocatalyst filter and the constituent of air cleaner with TiO_2 photocatalysts.
Reproduced from ref. 18 (*Zukai Hikarishokubai no Subete*, 2012) with permission from Ohmsha Ltd.

two photocatalytic TiO$_2$ filters with adsorbents are placed. The UV lamp placed between the two filters is continuously operated to decompose bacteria, viruses, and odor molecules photocatalytically. On deodorizing with an air cleaner, several kinds of adsorbent (active carbon and inorganic adsorbents) are often simultaneously utilized. The odor molecules in the air can be captured by these adsorbents, but as the surface is covered with odor molecules the performance declines. When the photocatalysts are placed close to the adsorbents, on irradiation of UV light, the odor molecules diffusing over the surface of the adsorbent are decomposed to carbon dioxide and water by contacting with photocatalysts and the concentration of the odor molecules decreases.

Agricultural products like vegetables and fruits continue life activities after harvest, causing cellular respiration. Therefore, during transportation, the freshness is lowered by the effects of the contamination of the microorganisms and ethylene gas emitted from the products. The air cleaner, which can be placed inside a container, is practically developed to eliminate the decomposing organic gas and bacteria.[18]

References

1. T. Saitoh, *Sankachitan Hikarishokubai no Subete (All of the TiO$_2$ Photocatalysis)*, ed. K. Hashimoto and A. Fujishima, CMC Press, Tokyo, 1998, p. 94.
2. M. Suzuki, *Kagaku Kogaku Ronbunshu*, 1994, **20**, 113.
3. M. Cargnello, T. R. Gordon and C. B. Murray, *Chem. Rev.*, 2014, **114**, 9319.
4. S. Sakka, *Zoru-geru Hou no Kagaku (Science of Sol-gel Method)*, Agune-Shofu-Sha, Tokyo, 1988.
5. C. J. Brinker and G. W. Scherer, *Sol-Gel Science*, Academic Press, London, 1990, ch. 13 (Film Formation), p. 786.
6. Y. Shigesato, *Zukai Hikarishokubai no Subete (Illustrated All of the Photocatalysis)*, ed. K. Hashimoto and A. Fujishima, Ohmusha Inc., Tokyo, 2012, p. 75.
7. C. B. Carter and M. G. Norton, *Ceramic Materials Science and Engineering*, Springer, New York, NY, 2007.
8. M. Akanuma, *Zukai Hikarishokubai no Subete (Illustrated All of the Photocatalysis)*, ed. K. Hashimoto and A. Fujishima, Ohmusha Inc., Tokyo, 2012, p. 78.
9. J. Tanaka, *Photocatalysis*, 2004, **13**, 6.

10. N. Nakayama and K. Takami, *Hikarishokubai (Photocatalyst)*, ed. K. Hashimoto, B. Ohtani and A. Kudo, NTS Books, Tokyo, 2005, p. 582.
11. J. Oguma, Y. Nosaka *et al.*, *Eur. Coat. J.*, 2011, **12**, 84.
12. T. Watanabe, *Sankachitan Hikarishokubai no Subete (All of the TiO$_2$ Photocatalysis)*, ed. K. Hashimoto and A. Fujishima, CMC Press, Tokyo, 1998, p. 203.
13. S. Fujisawa, *Kinou Zairyou*, CMC Publishing, Tokyo, 2003, vol. 23, p. 57.
14. H. Toyoda, *Zukai Hikarishokubai no Subete (Illustrated All of the Photocatalysis)*, ed. K. Hashimoto and A. Fujishima, Ohmusha Inc., Tokyo, 2012, p. 134.
15. H. Fukui, *Zukai Hikarishokubai no Subete (Illustrated All of the Photocatalysis)*, ed. K. Hashimoto and A. Fujishima, Ohmusha Inc., Tokyo, 2012, p. 140.
16. N. Tanaka, *Zukai Hikarishokubai no Subete (Illustrated All of the Photocatalysis)*, ed. K. Hashimoto and A. Fujishima, Ohmusha Inc., Tokyo, 2012, p. 167.
17. H. Matsubara *et al.*, *Rep. Gifu Prefect. Pap. Res. Inst.*, 1995, 7.
18. H. Ando, *Zukai Hikarishokubai no Subete (Illustrated All of the Photocatalysis)*, ed. K. Hashimoto and A. Fujishima, Ohmusha Inc., Tokyo, 2012, p. 216.

9 Evaluation Methods of Reactivity

9.1 Outline of the Standard Tests

ISO (the International Organization for Standardization) assists manufacturers to develop and deliver products because the products of photocatalysts for environmental use are developed in industries. The products should have the defined characteristics desired by their customers such as activity, robustness, appearance. Thus, for industry these standards ensure their products are widely accepted and competitive, whereas for consumers they ensure product quality and reliability. Although the ISO has no legal authority to enforce the implementation of its standards, it is worth noting that countries sometimes choose to refer to them in regulative publication.[1]

Technical Committee ISO/TC 206 (Fine ceramics)[2] publishes the international standards for the test of photocatalysts. The standards of the product inspection were proposed initially for UV light responsive photocatalysts and successively for visible light responsive photocatalysts taking into the account the indoor use, which are summarized in Table 9.1.

For ISO as indicated in Table 9.1 besides the standards of the light sources used for the performance tests of photocatalysts, the testing issues are categorized into self-cleaning, air-purification, and water-purification which have been successively standardized since 2007 or under the consideration as ISO standard. Because the productization of photocatalysts began in Japan, Japanese Industrial Standard (JIS) has been published successively since 2004. Then, the testing

Introduction to Photocatalysis: From Basic Science to Applications
By Yoshio Nosaka and Atsuko Nosaka
© Yoshio Nosaka and Atsuko Nosaka, 2016
Published by the Royal Society of Chemistry, www.rsc.org

Table 9.1 Summary of existing, forthcoming, and suggested ISO[2] and JIS tests in 2015.[a]

	Test target	Ultraviolet (UV) light		Indoor light (Vis)	
		ISO	JIS	ISO	JIS
Light source		ISO 10677	R1709	ISO 14605	R1750
Self-cleaning	Contact angle	ISO 27448	R1703-1	CD 19810	R1753
	Methylene blue	ISO 10678	R1703-2		
Air purification	Nitric oxide	ISO 22197-1	R1701-1	CD 17168-1	R1751-1
	Acetaldehyde	ISO 22197-2	R1701-2	CD 17168-2	R1751-2
	Toluene	ISO 22197-3	R1701-3	CD 17168-3	R1751-3
	Formaldehyde	ISO 22197-4	R1701-4	CD 17168-4	R1751-4
	Methylmercaptane	ISO 22197-5	R1701-5	CD 17168-5	R1751-5
	Test chamber			ISO 18560-1	R1751-6
Water purification	DMSO	ISO 10676	R1704		
Disinfection	Bacteria	ISO 27447	R1702	ISO 17094	R1752
	Fungi	ISO 13125	R1705		
	Viruses	ISO 18061	R1706	CD 18071	R1756
	Algae	CD 19635		AWI 19652	R1757
Others	Complete decomposition		R1708		
	Dissolved oxygen	CD 19722			
	Quantum efficiency	AWI 19728			

[a]DMSO: Dimethyl sulfoxide, CD: Committee draft, AWI: Approved work item.

methods of JIS corresponding to those of ISO have been added to Table 9.1 as reference. The individual issues will be described in the following sections.

Photocatalysts are usually immobilized on the flat or porous substrate of the products. Therefore, tests of such products are mainly for ISO. However, because research on photocatalysts is mostly performed in powder form, adequate test methods for the activity of powders are partly under consideration for ISO. Namely, tests on the complete decomposition of acetaldehyde (see Section 9.5) and the removal of dissolved oxygen (see Section 9.7) are actually under consideration.

For water decomposition, ISO does not consider the publication because the industrialization for water decomposition by photocatalysts has not yet been established. However, because considerable numbers of research reports have been published, the examples on the tests of activities of water decomposition with powder photocatalysts will be introduced at the end of this chapter.

9.2 Light Sources

9.2.1 For Ultraviolet (UV) Irradiation Test

With a UV fluorescent lamp various spectral distributions can be obtained by combining the fluorescent substances and lamp bulbs.[1] The UV lamp using glass through which UV light passes but which absorbs visible light is called black light blue (BLB). On the other hand, in the case of usual UV fluorescent bulbs, there appear also bright lines in the visible region and the lamp is called a black light (BL).

Four kinds of UV fluorescent lamps (351BLB/BL, 368BLB/BL) and a Xenon lamp are prescribed by ISO 10677 as test light sources for UV responsive photocatalysts. The wavelength profiles for two kinds of BLB lamps and a Xenon lamp are shown in Figure 9.1.[1] The numbers 351 and 368 represent the wavelength of the emission bands which are attributed to the difference of the phosphor for lead-doped barium silicate and europium-doped fluoroborate, respectively.[1]

ISO admits the usage of the UV light regions of a xenon lamp possessing the distribution of the wavelength similar to that of sunlight supposing the usage of the UV light region of the sunlight. The standard also recommends the use of a photometer to measure the UV irradiance. Measurements of UV irradiance should start at

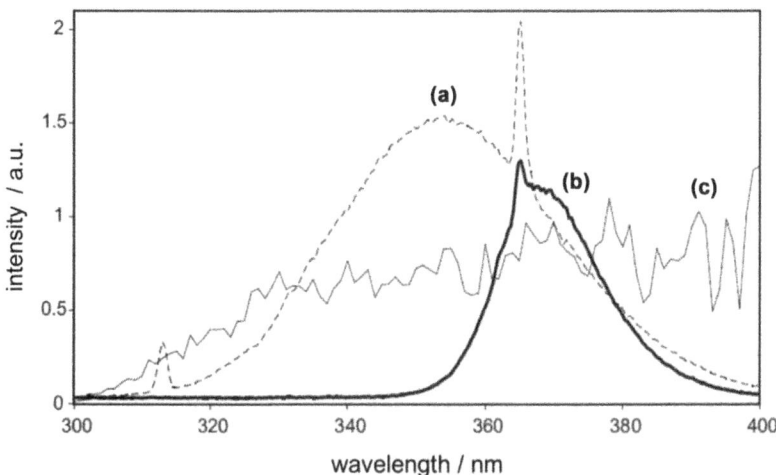

Figure 9.1 Wavelength profiles of relative light intensity for BLB light source centered at (a) 351 and (b) 368 nm, and (c) a xenon arc lamp.
Reprinted from A. Mills, C. Hill, and P. K. J. Robertson, Overview of the current ISO tests for photocatalytic materials, *J. Photochem. Photobiol. A: Chem.*, **237**, 7–23. Copyright (2012) with permission from Elsevier.

least 15 min. after the lamp is switched on and be performed also at the start and at the end of the test period.

9.2.2 For Indoor Lighting Test

In the literature on photocatalysts, visible light often means the light of wavelength between 400 and 700 nm. However, as a term in electronics it is widely defined that the wavelength of visible light is between 360 and 830 nm. Therefore, ISO defines that visible light responsive photocatalysts are those used under room light and that the corresponding light sources are fluorescent light lamps, which are often used as room lights.

The typical irradiation spectrum of a fluorescent lamp using halophosphate salt as a phosphor was already shown in Figure 2.12. ISO admits a three wavelengths (triphosphor) fluorescent lamp. Anyway, because not only visible light but also the bright lines of mercury are contained in the UV region ($\lambda < 400$ nm), the use of a filter is advocated. When a filter cutting the wavelength shorter than 416 nm is used the lighting is defined as Type A, that shorter than 392 nm as Type B, and without filter is defined as Type C, in which the filter to be used is designated by ISO 14605. For Type A the filter which passes light of wavelength of longer than 400 nm (transmittance is less than 0.1% at 400 nm) and for Type B the sharp cut filter which passes light of wavelength of longer than 360 nm are used, respectively.[3]

As the unit of the intensity of visible light, that is the illuminance, generally lux (lx) is used. 6000 lx is usually adopted as the maximum illuminance, which can be easily attained.[4] Because lux expresses the light intensity perceived with the human eye, the conversion efficiency with the light intensity ($W\ m^{-2}$) is different depending on the wavelength. The radiation in the wavelength range of 400–700 nm with white fluorescent lighting of 1000 lx corresponds to 0.29 $mW\ cm^{-2}$.[5] Therefore, the power of white fluorescent lighting of 6000 lx corresponds to 1.7 $mW\ cm^{-2}$. Comparing with sunlight, the illuminance of sunlight at the ground surface (air mass 1.5) is 110 000 lx, which is about 44 $mW\ cm^{-2}$ in 400–700 nm.[5]

About 5 mg of mercury is used for one fluorescent lamp, which can affect the environment at the time of the disposal, and the energy efficiency is not as good as that of an LED (light emitting diode). Hence, the white LED is now being selected instead of the fluorescent lamp. Because the BLB and fluorescent lamp will become difficult to purchase in the near future, reassessment of the light sources will be needed.

9.3   Self-cleaning

9.3.1   Decomposition of Methylene Blue in Solution

Because the concentration of dye can be easily measured, in the literature the decomposition of dye is often reported as a test of the activity performance. However, one must be careful because dye absorbs irradiation light. The decomposition of dye prescribed by ISO is provided by the irradiation of UV-light only and methylene blue (MB) is used as dye. As shown in Figure 9.2, the absorption spectrum of MB characteristically exhibits the large molar absorptivity at 664 nm but almost no absorption at 360–380 nm,[6] which is the wavelength for BLB light, nor at around 450 nm, which is the wavelength for LED.

The apparatus shown in Figure 9.3 is used for decomposition of MB, where the film is pressed against the bottom of the cylinder or the film can be just placed at the bottom of the cylinder. The cylinder is made of resin or glass whose inside diameter and height are 40 mm and 30 mm, respectively. The conditioning is conducted as follows: by pouring 35 mL MB solution of 2×10^{-5} M into the cylinder, and leaving to stand for 12 h to reach the sufficient adsorption of MB.

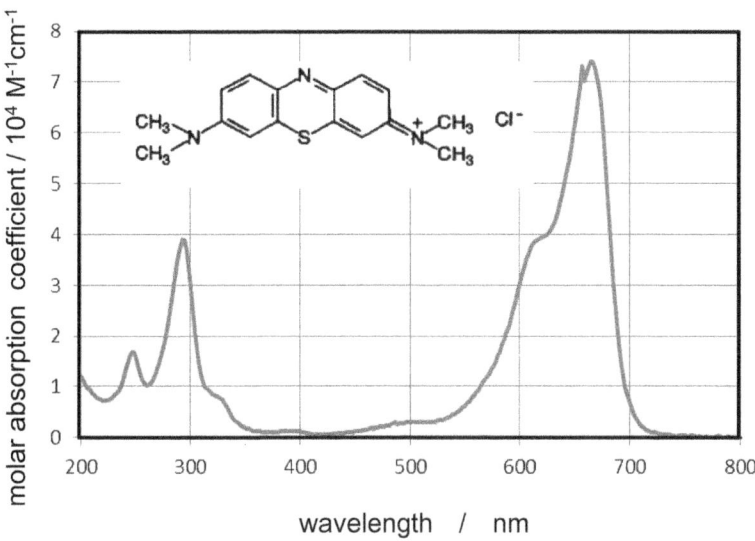

Figure 9.2 Chemical structure and the absorption spectrum of methylene blue in aqueous solution. $M = mol\ dm^{-3}$.
Data were taken from ref. 6.

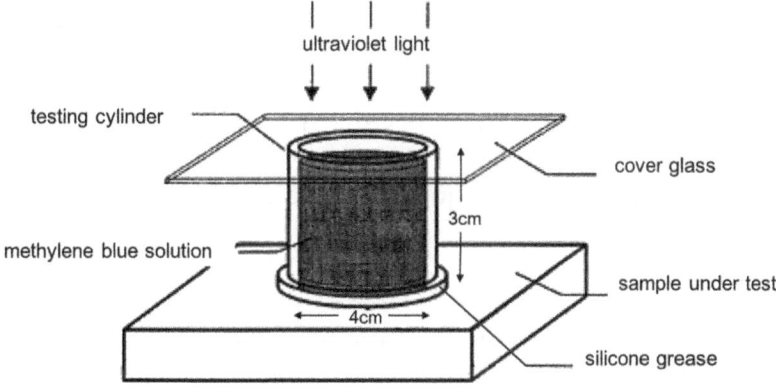

Figure 9.3 Irradiation set-up for the determination of photocatalytic activity of surfaces in an aqueous solution by degradation of methylene blue.

Next, 35 mL MB solution of 1×10^{-5} M is poured in the cylinder, and UV light is irradiated from the top of the surface for 20 min. After sampling the solution for absorption measurement, the solution is returned to the cylinder with stirring for the next irradiation. The operations of 20 min. irradiation and measurements of absorbance are repeated at least nine times. For three samples the above experiments are performed and a decomposition activity index, r, is calculated by averaging the decrease rates of the MB concentration $(nM\ min^{-1})$.

Let's calculate the relation of the activity index r (or the decrease rates of the MB concentration in $nM\ min^{-1}$) with the apparent quantum yield AQY (%). For the representative wavelength λ (368 nm) of UV light $I (= 1\ mW\ cm^{-2})$, as indicated in Section 2.1, the number of irradiated photons can be calculated by multiplying the light intensity I and wavelength λ with the constant $(1/N_A ch)$. That is, the photon flux becomes $3.08 \times 10^{-9} (= 8.36 \times 368 \times 10^{-12})\ mol\ cm^{-2}\ s^{-1}$. Because the area of the irradiation is $\pi \times 2.0^2\ cm^2$, the apparent quantum yield becomes $AQY = (r \times 10^{-9} \times 0.035/60) \times 100/(\pi \times 2.0^2 \times 3.08 \times 10^{-9}) = 0.00152 \times r$ (%).

In the experiment for MB of 10^{-5} M $(= 10^4$ nM), the initial absorbance is about 0.74, which decreases with decreasing the absorbance by ΔAbs after 20 min. From the value of ΔAbs, r is $(\Delta Abs/0.74) \times 10^4$ nM/20 min. $= 676 \times \Delta Abs$ $(nM\ min^{-1})$. Therefore, AQY is estimated to be $1.03 \times \Delta Abs$ (%). Namely, when the absorbance decreases by 0.01 in the reaction for 20 min. in this test, the apparent quantum yield of the MB decomposition is estimated to be about 0.01%.

However, it must be noticed that the decrease of the absorbance at 664 nm occurs at the very initial step of the decomposition of MB. Photocatalytically, MB may be completely decomposed as follows.[7]

$$C_{16}H_{18}N_3SCl + 25.5O_2 \rightarrow HCl + H_2SO_4 + 3HNO_3 + 16CO_2 + 6H_2O \quad (9.1)$$

Thus, a lot of photons are required for the complete decomposition.

9.3.2 Decrease of Water Contact Angle

When an oil stain is adsorbed on the test board coated with photo-catalyst, when water drops are added the contact angle of the water against the surface becomes large. However, along with the photo-catalytic decomposition of the oil stain, the contact angle decreases. Hence, the rate of the decrease of contact angle was adopted to evaluate the self-cleaning effect of photocatalysts. Because the materials for outdoors which utilize UV light were initially developed, the evaluation method using UV light as a light source (ISO 27448) was firstly published, then the evaluation method under room light conditions (ISO 19810) was published. For measuring the contact angle, the surface must be flat. Therefore, this evaluation method cannot be applied for water permeable samples, which cannot hold water drops. The correlation of the results using the various photocatalytic materials by this method with those of the activity evaluation by MB decomposition is described in JIS R1703-1.

There are several methods to measure the contact angle. For a flat sample with a width of more than several cm, it is convenient and highly accurate to use a tilting plate method, in which part of the flat sample is immersed aslant and the angle at which a meniscus cannot be formed is measured as a contact angle.

Commercial contact angle gauges often employ a drop method. For the drop method as shown in Figure 9.4A, the contact angle is the angle of the tangent line of the water droplet against the solid surface at the edge of the water. The angle is measured with a protractor or as the angle of reflection of laser light. Otherwise, with the expanse (w) and the height (h) of the droplet the contact angle θ can be also calculated by using eqn (9.2).

$$\theta = 2 \tan^{-1}(2h/w) \tag{9.2}$$

When the size of the droplet is extremely small and it is not affected by gravity, the droplet can be regarded as a part of a sphere. Therefore, the height can also be calculated from the volume of the droplet.

One must be cautious about the evaporation of water during the measurements. On dropping, the contact edge advances and then sweeps back when the volume is decreased by evaporation. When the difference between the advance and the sweepback of the contact angle is large, accurate measurement of the contact angle is difficult. When the surface is not flat and the real surface area is r times larger than the apparent surface area, the measured contact angle θ' holds the relationship $\cos \theta' = r \times \cos \theta$. Furthermore, it is desirable that measurements are performed on several places on the surface of the same solid materials to assure the same results can be obtained.

As for the evaluation index, for the evaluation method adopting UV light, it is evaluated how far the contact angle falls on the UV irradiation, namely evaluated by the final contact angle. On the other hand, for the evaluation method under room light conditions, as shown in Figure 9.4B, it is evaluated by the time at which the contact angle decreases down to a half or one-tenth of the initial value.

For evaluation of the activity of photocatalysts, firstly the pretreatments to remove the stain by the irradiation of UV light are conducted to make the contact angle less than 10 degrees. When the contact angle does not become less than 10 degrees, this evaluation method cannot be applied to the sample. Afterwards, stearic acid dissolved in heptane solution is coated by spin coating and dried. For the initial

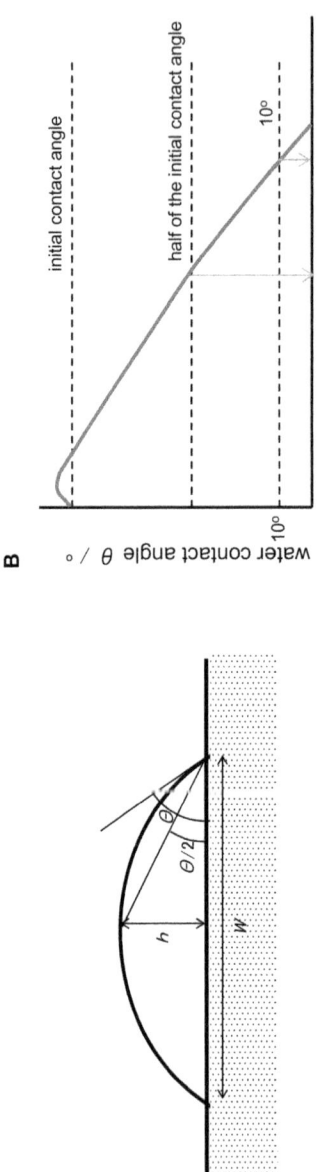

Figure 9.4 (A) Contact angle of water droplet and (B) evaluation indices.

evaluation method for UV irradiation, the pretreatment was not performed, as a coating substance oleic acid was used instead of stearic acid, and the method other than spin coating method was assigned as a coating method.

9.3.3 Simple Method with a Testing Ink

A simple and inexpensive method for the rapid testing of the photocatalytic activity of self-cleaning surfaces was proposed.[8] The use of an inexpensive digital scanner provides all the advantages of using UV/ Visible spectrophotometry in recording the change in color of ink, without need to use expensive, bulky equipment or significant technical support.

The photocatalyst indicator, Rz ink, consists of resazurin (Rz) dye solution containing hydroxyethyl-cellulose, glycerol, and a surfactant. This ink is coated on the sample and, after drying at 70 °C for 10 min., the sample is irradiated with BLB. Consequently, Rz dye is reduced to resorufin (Rf), and the color changes from blue to pink as the spectrum changes shown in Figure 9.5.[9] On the other hand, glycerol is oxidized

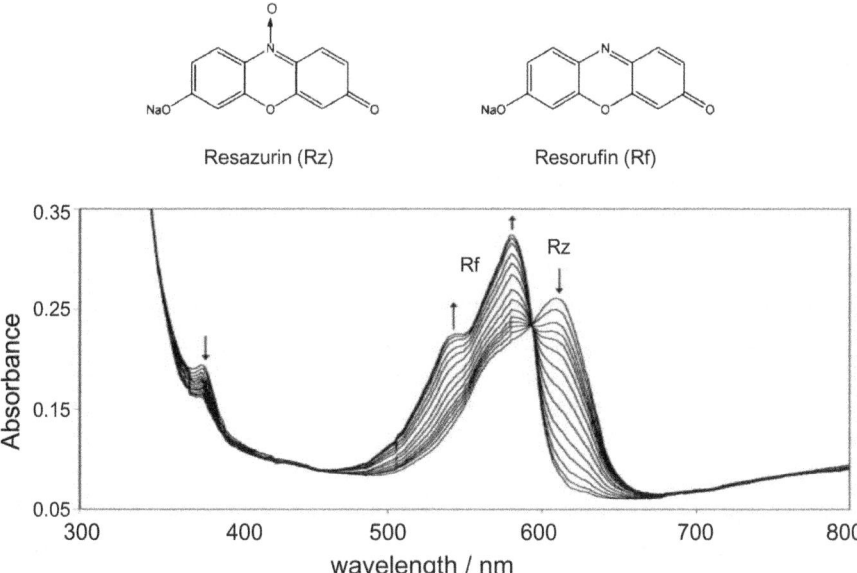

Figure 9.5 UV/Visible absorption spectra of a sample glass coated with resazurin/glycerol/hydroxyethyl cellulose ink and the effect of UV (BLB) irradiation. The spectra were recorded every 30 s. Reproduced from ref. 9 with permission from the Royal Society of Chemistry.

as a sacrificial electron donor by the photocatalyst. Because only the absorption above 600 nm is extinguished, the sample surface is recorded with a handheld document scanner, and the increment of the R component in RGB colors corresponds to the photocatalytic activity. The inter-laboratory repeatability of the Rz ink test is high (distribution was *ca.* 8.1%) and better than many of the current ISO photocatalyst tests.[8] As a consequence, the above method appears particularly suitable for measuring the photocatalytic activity of self-cleaning glass paints and tiles both in the lab and in the field, at small cost and with a little training, but with a reasonable degree of precision.[8]

9.4   Air Purification

Though the applications of photocatalysts are highly diversified, the first test method published by ISO was the removal of nitric oxide (NO) by UV light (ISO 22197-1), and the expectation on the air purification is high.[10]

Later, the evaluation test method for various VOC (Volatile Organic Compounds) with a similar apparatus was published. However, because of the limitation to remove VOC with the apparatus of the flow-through type targeting outdoor use by flat photocatalysts, the methods using a reactor of small chamber type are under consideration (ISO/CD 17168-1).

9.4.1   Removal of Nitric Oxide

A schematic drawing of the apparatus for an air purification test is shown in Figure 9.6A. Under room light conditions, a filter to remove UV light is set on the fluorescent lamp. The size of the test sample is $10 \text{ cm} \times 5 \text{ cm}$. The test gas diluted with air of humidity 50% is flowed through the 5 mm thick space above the surface of the sample which is held to make no step. When the test sample is a porous honeycomb shape, the measurements are conducted in the arrangement as shown in Figure 9.6B.

In the case of the removal test of NO, for the detection of NO and NO_2 a chemiluminescence detector is used while for the detection of NO_3^- ion chromatography is utilized. The concentration of NO gas is 1 ppmv $(= 1 \text{ } \mu\text{L L}^{-1})$ and flow velocity f is 3.0 L min^{-1}. An example for the measurement is shown in Figure 9.7, where the hatched area A indicates the amount of adsorption and the hatched areas B and C are the photoremoved amounts of NO and the generated amount of NO_2,

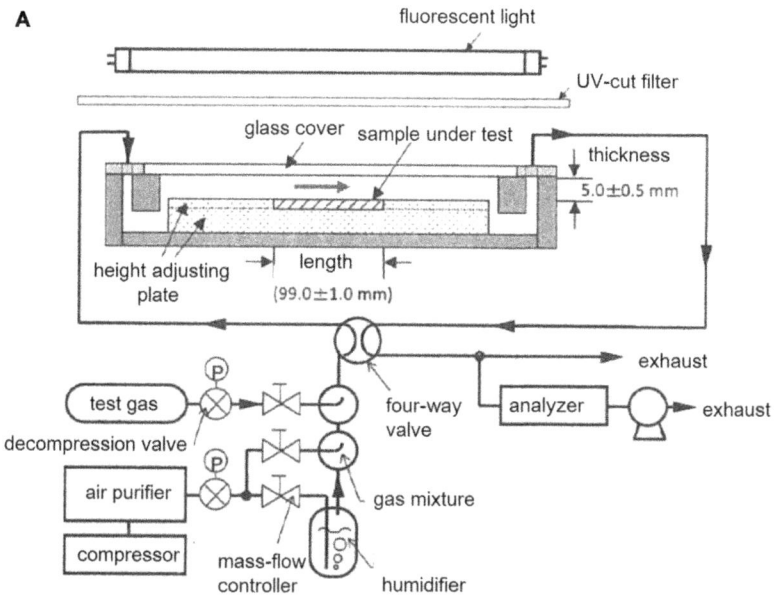

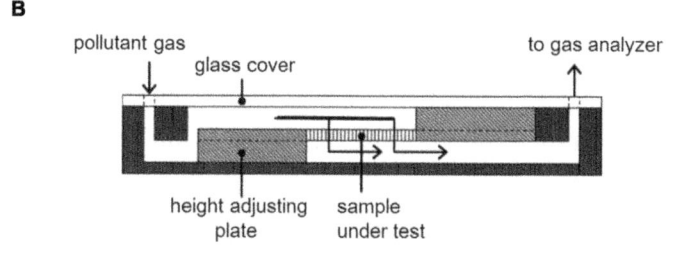

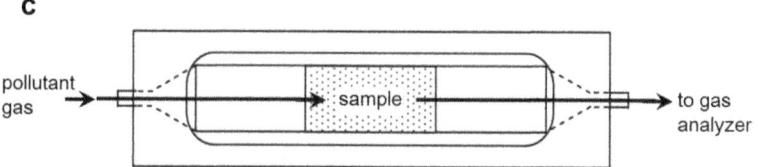

Figure 9.6 (A) Irradiation set-up for air purification test. (B) For test piece of filters. (C) Top view of the sample holder.
Reprinted from A. Mills, C. Hill, and P. K. J. Robertson, Overview of the current ISO tests for photocatalytic materials, *J. Photochem. Photobiol. A: Chem.*, **237**, 7–23. Copyright (2012) with permission from Elsevier.

respectively. The net removed amount of NOx by the test piece, n_{NOx}, is as follows:

$$n_{Nox} = \left(\frac{f \times 60}{22.4}\right)(B - C) \tag{9.3}$$

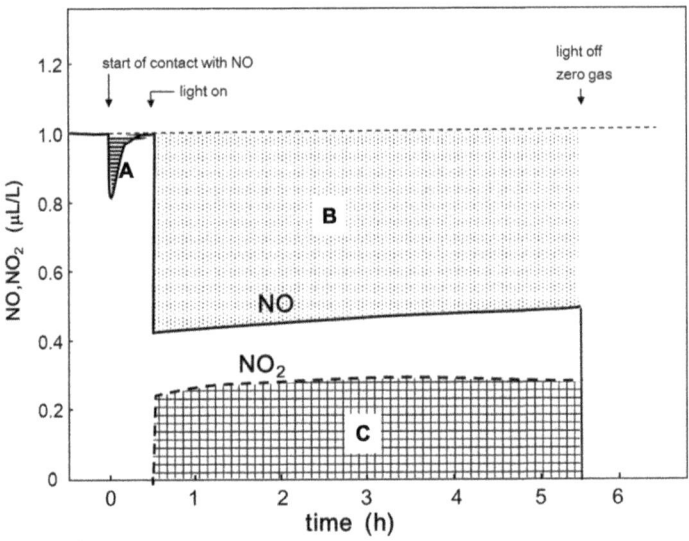

Figure 9.7 Typical data set generated in an NO removal test.
Reprinted from A. Mills, C. Hill and P. K. J. Robertson, Overview
of the current ISO tests for photocatalytic materials, *J. Photochem. Photobiol. A: Chem.*, **237**, 7–23. Copyright (2012) with
permission from Elsevier.

For the photocatalytic decomposition of NO, the following three
reactions take place.[8]

$$4NO + O_2 + 2H_2O \rightarrow 4HNO_2 \tag{9.4}$$

$$2HNO_2 + O_2 \rightarrow 2HNO_3 \tag{9.5}$$

$$2HNO_3 + NO \rightarrow 3NO_2 + H_2O \tag{9.6}$$

Therefore, the removal rate of NO matches with the generation rate of
NO_2. The detection of nitric acid HNO_3 is also required, which corresponds to the difference between the reduced amount of NO and
the generated amount of NO_2.

9.4.2 Removal of Organic Pollutants

Evaluation tests on removing VOCs (acetaldehyde, toluene, formaldehyde, and methylmercaptane) by using the products of photocatalysts are performed with a test piece holder as shown in
Figure 9.6. For the analysis of organic compounds on using gas
chromatography (GC), regular sampling is conducted with a 6-way

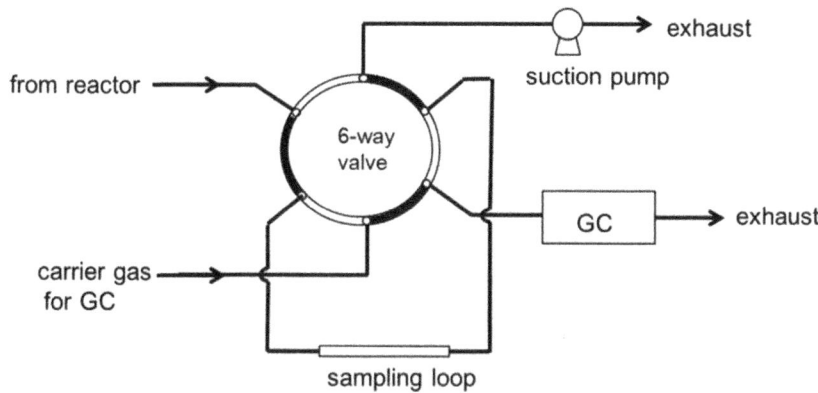

Figure 9.8 Analyzing unit for detection with gas chromatography (GC).

valve as shown in Figure 9.8. Otherwise, a gas-tight syringe with high reproducibility can be used. As for the removal of acetaldehyde (ISO 22197-2), the detection can be performed with GC, in-line FTIR, or non-dispersive IR spectrometer, and continuous measurements of the concentration are possible.

The removal of organic molecules for the test sample is conducted by the irradiation of UV light for 12–24 h. Then the experiments are initiated after ensuring that the concentration of CO_2 is less than 1 ppmv (1 μL L^{-1}) when the light source is irradiated under the flow of zero gas (synthetic air without CO_2) with 50% humidity. The measurement of CO_2 is not conducted when CO_2 does not become less than 1 ppmv due to the decomposition of binders in the sample. Under dark conditions, a test glass containing acetaldehyde of 5 ppmv is supplied at the flow rate of $f=1.0$ mL min^{-1} for adsorption. When the adsorption of acetaldehyde still continues 90 min. later and the concentration does not exceed 90% of the supplied concentration, the test method is judged inadequate for this test sample.

A typical experimental example is shown in Figure 9.9,[1] where the concentrations of acetaldehyde and CO_2 are expressed with solid and dotted lines, respectively. Photoirradiation is conducted for 3 h, starting from 30 min. after the initiation of the test. With the area B which corresponds to the decrease of the acetaldehyde from 5 ppmv during the photoirradiation, the removal of acetaldehyde n_A can be calculated with the following equation.

$$n_A = \left(\frac{f \times 60}{22.4}\right) B \tag{9.7}$$

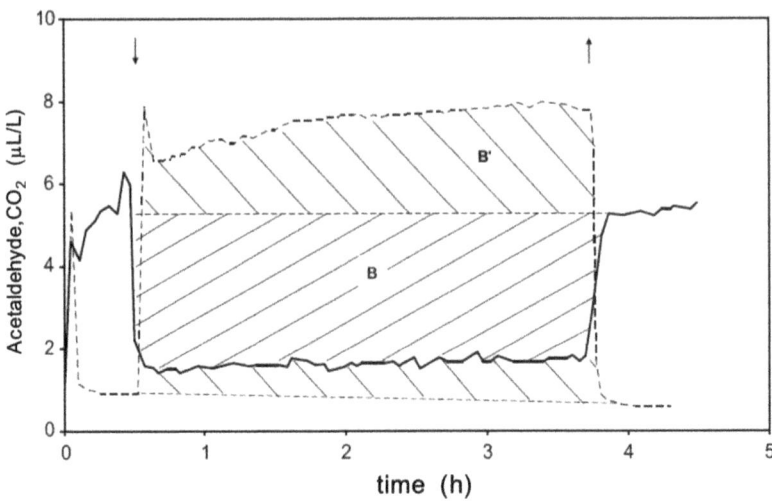

Figure 9.9 Typical data set for the ISO standard test for removal of acetaldehyde with generation of carbon dioxide.
Reprinted from A. Mills, C. Hill and P. K. J. Robertson, Overview
of the current ISO tests for photocatalytic materials, *J. Photochem. Photobiol. A: Chem.*, **237**, 7–23. Copyright (2012) with
permission from Elsevier.

The removal fraction $R\%$ is calculated with the following eqn (9.8)
with the concentration of acetaldehyde of [A] (ppmv) during the last
1 h and the supplied concentration of acetaldehyde of $[A]_0$.

$$R(\%) = \frac{[A]_0 - [A]}{[A]_0} \times 100 \qquad (9.8)$$

In Figure 9.9 the area B′, which is obtained by integrating the
concentration of CO_2, is about twice as large as that of B. For the
photocatalytic decomposition of acetaldehyde, acetic acid, formic
acid, and formaldehyde exist as intermediates as follows.[1]

$$CH_3CHO + 1/2\,O_2 \rightarrow CH_3COOH \qquad (9.9)$$

$$CH_3COOH + 1/2\,O_2 \rightarrow HCOOH + HCHO \qquad (9.10)$$

$$HCOOH + HCHO + 3/2\,O_2 \rightarrow 2CO_2 + 2H_2O \qquad (9.11)$$

Therefore, CO_2 of twice the amount of decomposed acetaldehyde is
generated as evidenced in Figure 9.9. Because CO_2 is generated

immediately after the decomposition of acetaldehyde, the decomposition rates of the intermediates are considered high.

For the removal test of toluene $C_6H_5CH_3$ (ISO 22197-3), $C_6H_5CH_3$ of the concentration of 1 ppmv is flowed. On decomposition, one molecule of toluene becomes $7CO_2$ *via* C_6H_5CHO and C_6H_5COOH. In the absence of water, these intermediates are adsorbed and the reactions become slow. Even in the presence of water the reactions of the intermediates for some photocatalysts become slow owing to the strong adsorption. Hence, the photocatalytic activity is not always the same as those obtained with the other methods.[1]

For the decomposition test of formaldehyde (HCHO) (ISO 22197-4), the humidified gas of 1 ppmv is utilized. The flow rate is usually $1.0 \, \text{mL min}^{-1}$ but when the removal rate is less than 5%, the flow rate of $0.5 \, \text{mL min}^{-1}$ or two test samples are adopted. For the detection, DNPH-HPLC (2,4-dinitrophenyl hydrazine-high performance liquid chromatography) method (for the case of JIS R1751-4) is adopted because GC cannot be used.

For the decomposition test of methylmercaptane (CH_3SH) (ISO 22197-5), the concentration of 5 ppmv and the flow rate of $1.0 \, \text{mL min}^{-1}$ are adopted. When the removal fraction is less than 5%, two pieces of test sample can be used. For JIS R1751-5, GC of FID is adopted for detection.

9.4.3 Removal of Formaldehyde by a Small Chamber Method

As a removal performance test of formaldehyde under room light irradiation, there is the above-mentioned method (ISO 22197-4) using the apparatus similar to that for the NO removal test. However, it is not applicable for the prediction of the effect under real environments. Then, a method using a test chamber which models real room environments was published as ISO 18560-1.

The concept of the apparatus is depicted in Figure 9.10. The gas flow rate is kept constant and the apparatus is constructed so as the gas of the inlet and the outlet do not mix. The backside and the edge of the test piece are covered with aluminum sheets so as not to contact with the gas. The sampling is conducted at the outlet and the inlet of gas. The background concentrations of total VOC and formaldehyde are not more than $2 \, \mu\text{g m}^{-3}$ and $20 \, \mu\text{g m}^{-3}$, respectively. The flow rate is controlled so that the mass transfer coefficient of water vapor becomes $15 \, \text{m h}^{-1}$. This roughly corresponds to the

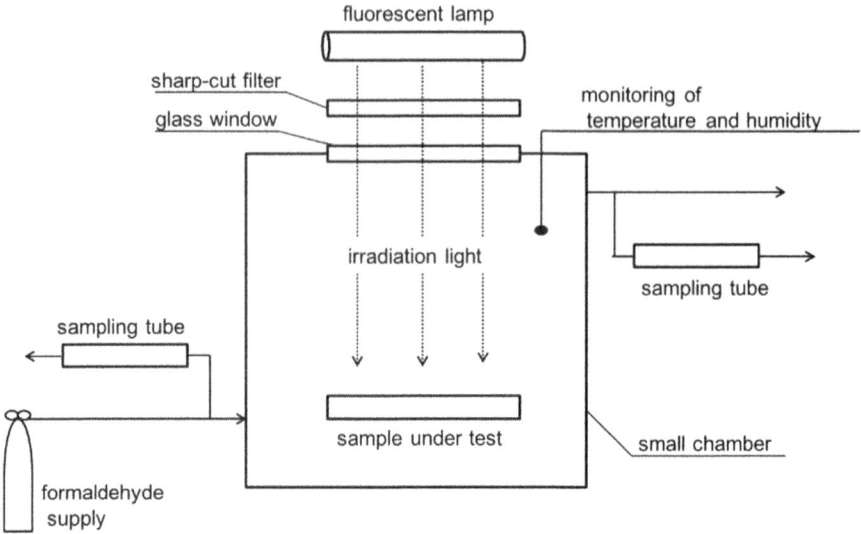

Figure 9.10 Conceptual diagram of a small chamber method (ISO 22197-4) for testing photocatalyst samples.

wind velocity of 0.25 m s^{-1} on the surface of the sample. The air exchange rate is 0.5 h^{-1} as a standard, and the illumination at the sample is 1000 lx by taking account of the real environment. Before flowing the gas, the test piece is cleaned with UV light irradiation. Then, the saturation in the adsorption of formaldehyde gas is confirmed by the fact that the concentration at the outlet becomes more than 90% of that at the inlet on taking more than one day. The concentration of formaldehyde obeys the guideline proposed by WHO, ρ_{gl}, which is 100 μg m^{-3} but in some cases it can be half or twice that amount.[11]

Afterwards, photo-illumination is conducted under the room light condition and the removal rate r (μg m^{-2} h^{-1}) is calculated with the ventilating volume of a small chamber, q_c(m^3 h^{-1}), and the surface area of test piece A (m^2) as follows,

$$r = (\rho_{in,t} - \rho_{out,t})q_c/A \tag{9.12}$$

where, $\rho_{in,t}$ and $\rho_{out,t}$ are the concentration (μg m^{-3}) of the supplied formaldehyde and that of the formaldehyde at the outlet at the passed time t, respectively. The removal rate r_{gl} (μg m^{-2} h^{-1}) at the concentration of guideline ρ_{gl} is calculated as follows.

$$r_{gl} = (\rho_{in,t}/\rho_{out,t} - 1)q_c \times \rho_{gl}/A \tag{9.13}$$

9.5 Complete Decomposition of Acetaldehyde

A test method for complete decomposition of acetaldehyde (ISO/AWI 19652), which is the evaluation method not for the commercial products but for the materials of photocatalysts to become the guide for the basic development, is proposed.

As shown in Figure 9.11, the powder sample is dispersed over the whole bottom of the petri dish of 6 cm diameter and placed in the closable vessel of 500 mL volume for photoirradiation. Then the stain on the photocatalysts is removed by loading zero gas $(O_2 : N_2 = 1 : 4)$ and irradiating with room light. It can be regarded as clean when the generation amount of CO_2 becomes no more than 2 ppm h^{-1}. The humidity in the reactor is set at 50% and the standard gas containing acetaldehyde of 100 ppm is injected. When it is confirmed that the adsorption is terminated when the decrease of acetaldehyde no longer takes place, the light of 10 000 lx is irradiated. The concentrations of acetaldehyde and CO_2 in the system are measured at every fixed time period. The measurements are completed when the concentration of CO_2 around 200 ppm is confirmed not to change any more as shown in Figure 9.11B. The detection of CO_2 is best performed with an FID-GC equipped with a methanizer. In addition, micro GC would be convenient if available.[12]

9.6 Water Purification with DMSO

Because the products of photocatalysts aiming at the decomposition of pollutants in water to purify are also developed, a test method for

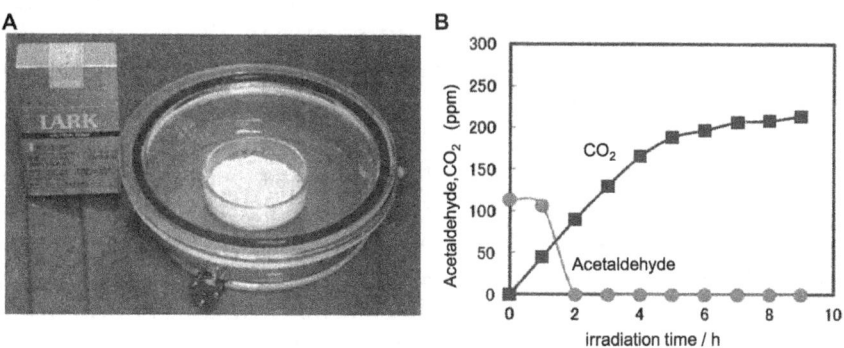

Figure 9.11 (A) Reaction chamber and (B) the example of the change of gas concentration for the test of complete decomposition of acetaldehyde.
Reproduced from ref. 12 (*Kaiho Hikarishokubai*, 2015) with permission from Photo Functionalized Materials Society.

water-purification performance of photocatalytic materials by measurement of forming ability of active oxygen is standardized (ISO 10676).

10 mg L^{-1} DMSO solution of 500 mL is circulated over the test piece of 10 cm×10 cm and the solution of 10 mL is collected at every fixed time period. Then, DMSO and produced MSA (methane sulfonic acid, CH_3SO_3H) in the solution are measured with GC or ion chromatography. This method is characteristic in terms of less adsorption as compared with the other methods. However, this method is not often utilized because DMSO is not a common pollutant, and ion chromatography is not an especially popular analytical instrument. It is suggested that instead of DMSO, phenol, which can be measured readily with spectroscopy, should be adopted.[1]

For this test method, DMSO is supposed to be photocatalytically decomposed to MSA owing to the exertion of active oxygen species, in particular OH radical. However, as stated in Chapter 6, photocatalytic reactions do not always proceed only with OH radicals. To observe OH radicals a phosphorescence spectroscopy would be simpler to operate, as stated in Chapter 5.

9.7   Consumption of Dissolved Oxygen

As an evaluation of the photocatalytic performance to remove organic pollutants in solution the decrease of TOC (Total Organic Carbon) can be measured. However, the photodecomposed intermediates of some organic compounds are sometimes too stable to be easily decomposed. Therefore, if TOC were standardized, applicable photocatalysts might be limited. Because oxygen is necessary for photocatalytic reactions a simpler test for photocatalytic activity to measure the decrease of the dissolved oxygen with an oxygen analyzer is proposed (ISO/CD 19722).

Because novel photocatalytic materials are often produced in powder form, the test method for the photocatalytic activity of powder materials is also required. The proposed method shown in Figure 9.12 can be applied for both thin film and powder samples as well.[13] The apparatus consists of a holder made of Teflon and a glass cylindrical vessel, the temperature in which is kept constant with a water-cooling jacket. The holder comprises a hole to hold an oxygen analyzer of diaphragm (thin film) type and a ditch hanging a test plate piece of effective surface of 3 cm×5 cm (right side drawing). For powder samples (left side drawing), the photocatalyst at the concentration of

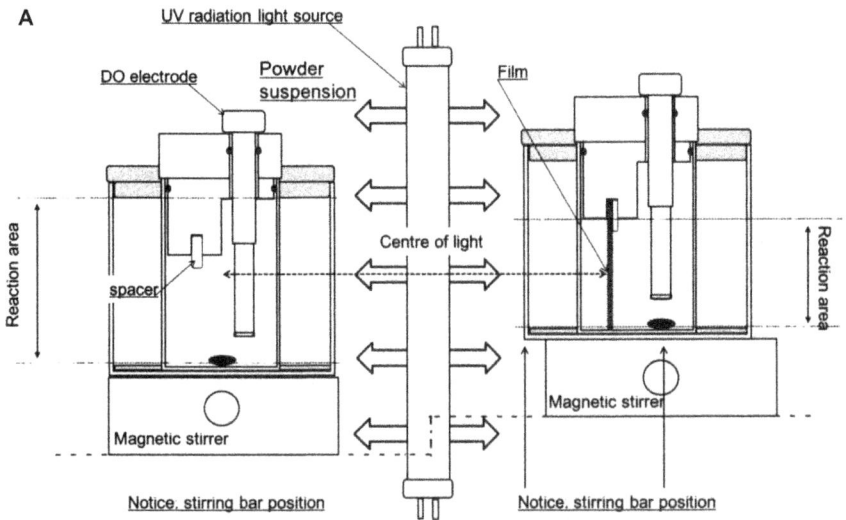

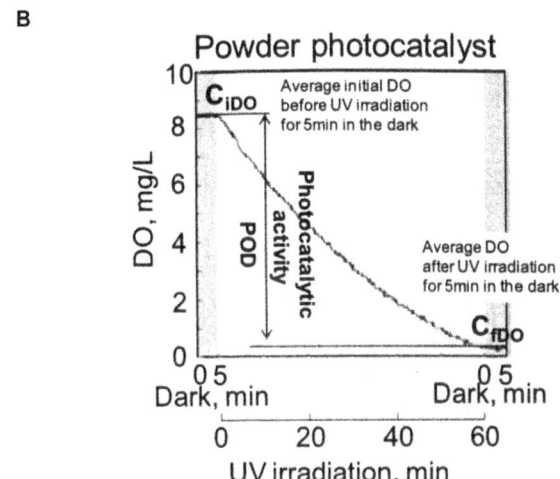

Figure 9.12   (A) A reaction cell for the test method of photocatalytic activity by dissolved oxygen consumption, and (B) a representative record sheet.
Reprinted from ref. 13 (*Kaiho Hikarishokubai*, 2015) with permission from Photo Functionalized Materials Society.

0.11 g L^{-1} is suspended as standard, the ditch of the plate is covered with a Teflon plate, and the irradiation position is shifted as indicated. The vessel is filled with 125 mL of 0.33 mM phenol solution so as to remove the air layer. To accelerate the diffusion of the dissolved oxygen in the solution, rapid stirring with a magnetic stirrer is necessary. $1.5 \, \text{mW cm}^{-2}$ is adopted as the standard for the intensity of the UV irradiation.

The example of the measurement is shown in Figure 9.12B. For the powder system, the activity is assessed by the period of time when the amount of the dissolved oxygen becomes zero or by the decrease amount of the dissolved oxygen after 60 min. photoirradiation. The activity of the photocatalytic plate materials with small surface area is assessed by the decreased amount of the dissolved oxygen after 180 min. irradiation.[13]

9.8 Disinfection

9.8.1 Antibacterial Activity

As the assessment methods for the antibacterial activity of photo-catalytic products, ISO 27447 for UV light and ISO 17094 for room light conditions were published. These can be applied to both film and plate products with antibacterial treatment but not to fiber products with antibacterial treatment.

After inoculating bacteria into the products with and those without photocatalytic treatment, the viable count after photoirradiation is measured. Then, the difference of the logarithm of the viable counts between the products with and those without photocatalytic treatment is used as an index of antibacterial activity. This value contains the viable bacterial count decreased without photoirradiation. For the antibacterial activity test, *Staphylococcus aureus* and *Escherichia coli* are adopted as bacteria, which can be obtained from the organizations which have joined the World Federation for Culture Collections.

For a film contact method, nine sample pieces (5 cm×5 cm, thickness of less than 10 mm) are prepared. Among them three pieces are used immediately after inoculating bacteria, three pieces are used after photoirradiation for a certain period of time, and three pieces are used after keeping in the dark for a certain period of time. On the other hand, when six test pieces with antibacterial treatment are prepared, three pieces are used after photoirradiation for a certain period of time, and three pieces are used after keeping in the dark for a certain period of time. As shown in Figure 9.13, a sterilized filter paper soaked with sterilized water for humidity control is placed on the bottom of the sterilized petri dish, and a glass pipe or a lid of a glass rod is placed so as not to contact the filter paper with the sample piece whose upside is the photocatalytic surface.

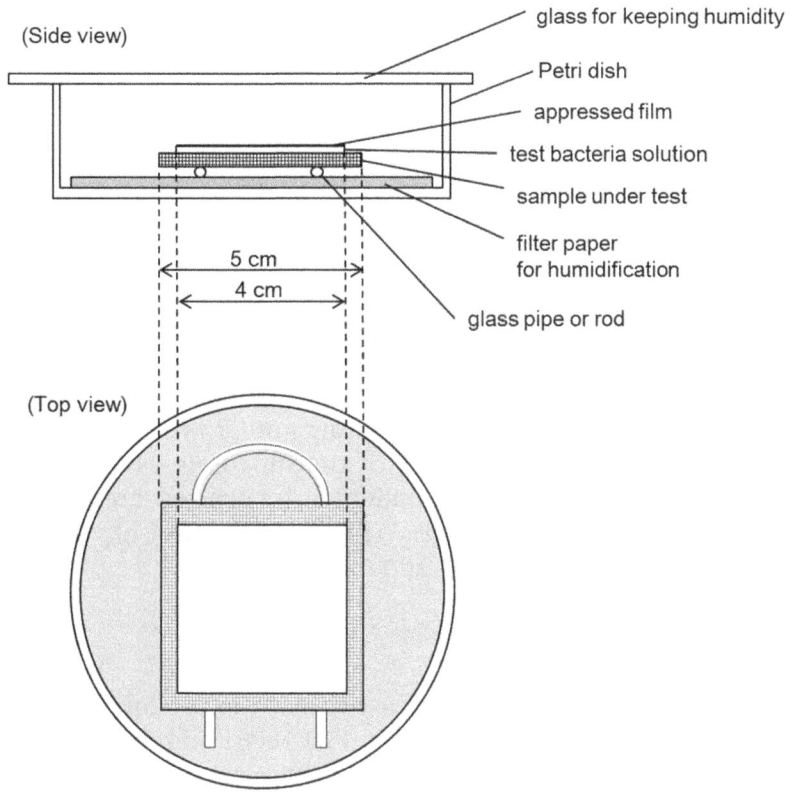

Figure 9.13 Set-up for antibacterial test with an appressed film.

The test bacteria solution of about 6.7×10^5 to $2.6\times10^6\,\text{mL}^{-1}$ is prepared and used immediately after the preparation. The bacteria solution of 0.15 mL is poured on the test piece which is covered with a sterilized coherence film (*e.g.* made from polypropylene, *etc.*), and then the petri dish is covered with a glass for humidity control. Photoirradiation is performed with the intensity of $0.1\,\text{mW cm}^{-2}$ for UV light source (in an indoor room close to the window in daytime) or $0.001\,\text{mW cm}^{-2}$ (an indoor place where sunlight does not enter or under UV light of a fluorescent light). In the case of indoor irradiation, the light intensity of 1000 lx is set as standard, while a maximum of 3000 lx is allowed. After being kept for 8 h (or 4 h) at 25 °C the test bacteria are washed out with a nutrient solution and a part of the solution is diluted by ten times. By counting the number of the colony after the cultivation of the diluted solution on the agar medium for two days, the original total number of viable bacteria can be estimated. The logarithm of the ratio of the total viable count to that for the sample without photocatalytic treatment is used to evaluate antibacterial activity.

9.8.2 Antifungal Activity

ISO 13125 defines a test method (UV light irradiation) of the antifungal activity of the products with photocatalytic treatment. For the suspension in which the spore of a fungus is homogeneously dispersed in physiological saline water containing surfactant, photoirradiation is conducted using the apparatus similar to that for the antibacterial activity test in Figure 9.13. The operation is similar to that for the antibacterial test stated above. By counting the number of spores of the fungus capable of sprouting and growth, the logarithm of the ratio of the viable number of the fungus is assigned to the photocatalytic activity. The fungi adopted for the test are *Aspergillus niger* and *Penicillium pinophilum.* The concentration of the fungal spore in the test solution is set at 5.0×10^5 mL^{-1}.

9.8.3 Antiviral Activity

The test methods of the antiviral activity of the products with photocatalytic treatment are described in ISO 18061 and 18071.

As shown in the outline in Figure 9.14 the test piece is vaccinated with bacteriophage Qβ, and the infectivity titer of the bacteriophage after photoirradiation is measured and assessed. Because the virus itself does not grow, by infecting *Escherichia coli*, the infectivity titer is assessed by counting the number of plaques of *Escherichia coli* on the agar medium. The concentration of the bacteriophage solution used for the test is prepared to be from 6.7×10^6 to 2.6×10^7 pfu mL^{-1}. The experiments are performed with the illuminance of no more than 200 lx for visible light and no more than 0.001 mW cm^{-2} for UV light.[14]

9.8.4 Anti-algal Activity

Microalgae can grow anywhere by producing nutritious substances by themselves when water and sun are supplied and play important roles as primary producers of the food chain. For instance, the symbiotic relationship of algae with fungi is well known. Algae not only spoil scenery by creating stains on the surfaces of buildings but they also provide fungi with nutrition, and erode paint and the inside of concrete to cause deterioration and corrosion. Hence, the anti-algal activity of photocatalysts is expected. However, because different from

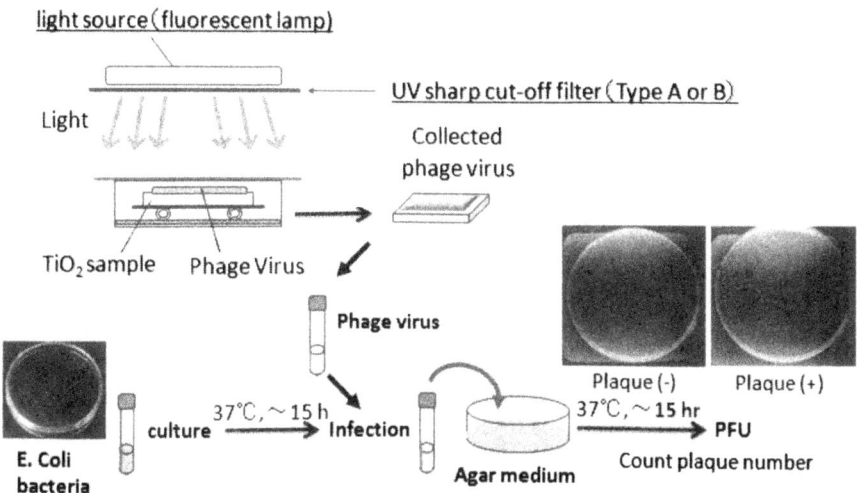

Figure 9.14 Procedures to test antiviral activity with indoor light. Reprinted from ref. 14 (*Kaiho Hikarishokubai*, 2015) with permission from Photo Functionalized Materials Society.

fungi it takes rather a long time to grow algae, the assessment methods by cultivation used for bacteria cannot be applied.[15]

Then, ISO/CD 19635 proposes a simple anti-algal activity test, using the principle that the chlorophyll pigment is faded when the cells of algae die out. In Figure 9.15A and B, the photomicrograph of *Chlorella vulgaris*, which is a kind of algae, and absorption spectrum of the solution of algae are shown, respectively. The optical absorbance at three representative wavelengths (660, 692, 750 nm) can disclose the absorption of chlorophyll and the intensities correspond to the quantity. Because this method to measure the quantity of chlorophyll from the absorption spectrum does not require any cultivation process to count the number of the algae in the solution, it can be performed easily as shown in Figure 9.15C.

9.9 Water Splitting

As an evaluation method for photocatalytic activity of water splitting, Kudo *et al.*, who published a number of reports on the complete decomposition of water, utilize a closed gas-circulation system equipped with a vacuum line.[16] A reaction cell and a gas sampling port are directly connected to a gas chromatograph (GC) as shown in Figure 9.16A. When photocatalytic activity is too high to use

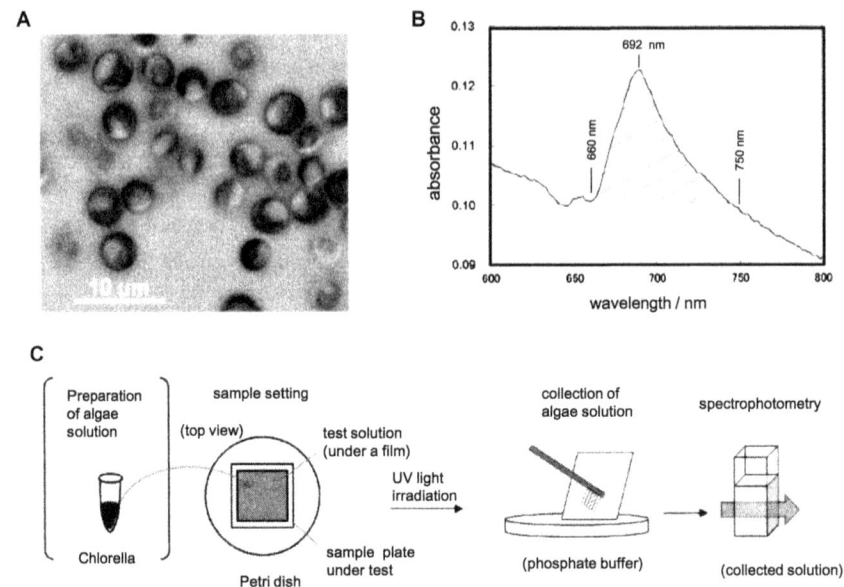

Figure 9.15 (A) Microscopic picture of *Chlorella vulgaris* and (B) the absorption spectrum of the solution. (C) Procedure to test for algae removal activity.
Reprinted from ref. 15 (*Kaiho Hikarishokubai*, 2015) with permission from Photo Functionalized Materials Society.

the GC, a volumetric method is used for the measurement of evolved gases.

The apparatus should be air-free because the detection of O_2 is very important for the evaluation of photocatalytic water splitting. There are several reaction cells. In general, efficient irradiation is conducted when an inner irradiation reaction cell is used. In Figure 9.16A an outer irradiation cell where the light is irradiated with a Xe lamp through a glass filter from above is shown. Briefly speaking, argon (Ar) gas is circulated with a pump through valves (b), (c), and (d). After photoirradiation for a certain period of time, the sampling of a part of the gas is conducted at the left side of a 3-way valve (a) and H_2 and O_2 are analyzed with GC.[16]

The important points on analysis of water decomposition reactions are (i) the ratio of generated H_2 with O_2 is $2:1$, (ii) the amount of the generated gas increases with the photoirradiation time, (iii) the dependency of the wavelength of the photoirradiation on the amount of the generated gas corresponds to that of the wavelength of the absorption, and (iv) the turn over number (TON) of the generated

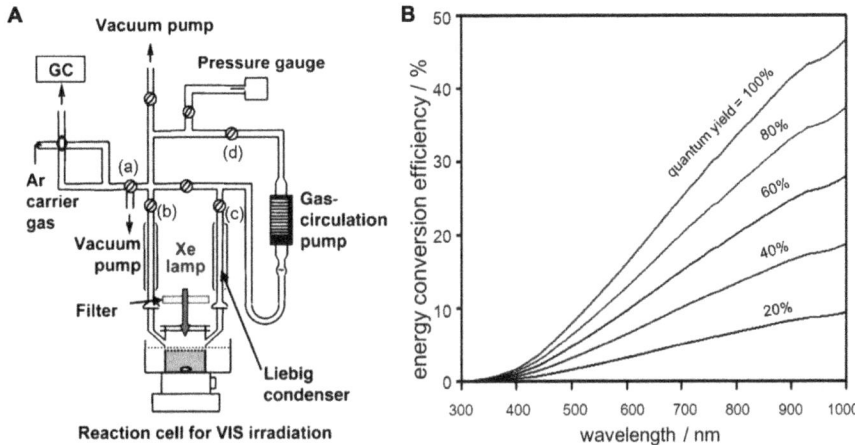

Figure 9.16 (A) Experimental set-up to test photocatalytic activity for water splitting. Reproduced from A. Kudo and Y. Miseki, *Chem. Soc. Rev.*, 2009, **38**, 253, with permission from the Royal Society of Chemistry.[16] (B) Calculated solar energy conversion efficiency as a function of wavelength for overall water splitting using photocatalysts with various quantum efficiencies. Reprinted with permission from K. Maeda and K. Domen, *J. Phys. Chem. Lett.*, 2010, **1**, 2655. Copyright 2010 American Chemical Society.[18]

amount is not less than 1, where the TON is the ratio of the generated amount of H_2 and O_2 to that of the active sites. Because the estimation of the active sites is generally difficult, the number of the surface atoms is sometimes used instead.[16]

In some literature, the photocatalytic activity is normalized by the amount of photocatalysts to express as the reaction rate per catalyst in the unit of mmol h^{-1} g^{-1}. Certainly, the amount of the absorbed photons increases proportionally with the amount of photocatalysts. However, it would not be appropriate because the difference of specific surface area or active sites is not taken into account. The generation amounts of gases such as H_2 and O_2 are expressed with the generation rate of gas (μmol h^{-1}) under the individual experimental conditions, and depend on the amount of the catalysts and the intensity of the photoirradiation. Hence the comparison of those obtained among different researches is difficult. Because the light intensity can be easily measured with a light power meter (photometer), the performance of the photocatalysts can be roughly compared by dividing the irradiated light intensity by the generation rate. This is called an apparent quantum yield (AQY) or incident photon to chemical conversion factor. Then, it is important to have knowledge about the wavelength dependence of the irradiated light source (*i.e.*, irradiance spectrum), the description of the optical filter, and also knowledge about the absorption spectra of the photocatalysts.

As stated in Chapter 2 (Section 2.5), a part of the incident light is absorbed in the semiconductor photocatalysts and the rest is reflected (scattered) or transmitted. Therefore, by taking account of the actual amount of the light absorption in the irradiated amount of light, the actual quantum yield can be estimated, and naturally it becomes larger than the AQY. To distinguish AQY from the actual quantum yield, terms such as quantum efficiency or photonic efficiency are recommended by the IUPAC (International Union of Pure and Applied Chemistry).[17]

Because the original purpose of the solar energy is to be utilized as hydrogen energy, it would be effective to estimate it with the solar to hydrogen conversion efficiency (STH). Because the power of the solar light is 112 mW cm^{-2}, STH is expressed by the following equation with the area of solar irradiation A (cm^2), and the generation rate of hydrogen r_H (mmol s^{-1}),

$$STH(\%) = 2 \times r_H \times F \times 1.23 \text{ V}/(A \times 112 \text{ mW cm}^{-2}) \times 100 \qquad (9.14)$$

where F is the Faraday constant. Then the term $r_H \times F$ has the unit of electric current (mA).

For STH (%) to correspond to the quantum yield, the relation between irradiance spectrum and the absorption spectrum becomes important. The spectrum of sunlight, as shown in Chapter 2 (Figure 2.12), has a vast expanse in the longer wavelength region. On the other hand, semiconductor photocatalysts can absorb only the light of shorter wavelength than the corresponding bandgap energy E_g ($\lambda = 1240/E_g$) and, moreover, the maximum energy that can be utilized is E_g. Therefore, even if the quantum efficiency were 100%, the available energy conversion would be altered as shown in Figure 9.16B. Because the quantum yield is actually less than 100%, the solar energy conversion decreases further as shown in the figure. Because the quantum yield for water decomposition so far is small, the research reports are now presented by AQY with limited excitation wavelength.

References

1. A. Mills, C. Hill and P. K. J. Robertson, *J. Photochem. Photobiol., A*, 2012, **237**, 7.

2. http://www.iso.org/iso/home/store/catalogue_tc/catalogue_tc_browse.htm?commid=54756.
3. S. Takeshita, *Photocatalysis (Gijutsu Kyouiku Shuppan, Tokyo)*, 2015, **46**, 58.
4. T. Sano and K. Takeuchi, *Photocatalysis (Gijutsu Kyouiku Shuppan, Tokyo)*, 2015, **46**, 76.
5. R. W. Thimijan and R. D. Heins, *HortScience*, 1983, **18**, 818.
6. Oregon Medical Laser Center. http://omlc.org/spectra/mb/mb-water.html.
7. R. W. Matthews, *J. Chem. Soc., Faraday Trans.*, 1989, **85**, 1291.
8. A. Mills *et al.*, *J. Photochem. Photobiol., A*, 2013, **272**, 18.
9. A. Mills *et al.*, *Chem. Commun.*, 2005, 2721.
10. A. Mills and S. Elouali, *J. Photochem. Photobiol., A*, 2015, **305**, 29.
11. Japan Industrial Standard, JIS R1751-6:2013.
12. Y. Kuroda, *Photocatalysis (Gijutsu Kyouiku Shuppan, Tokyo)*, 2015, **46**, 72.
13. T. Hirakawa, *Photocatalysis (Gijutsu Kyouiku Shuppan, Tokyo)*, 2015, **46**, 88.
14. A. Takeshita, H. Ishiguro and Y. Kubota, *Photocatalysis (Gijutsu Kyouiku Shuppan, Tokyo)*, 2015, **46**, 82.
15. Y. Ohko and T. Harada, *Photocatalysis (Gijutsu Kyouiku Shuppan, Tokyo)*, 2015, **46**, 94.
16. A. Kudo and Y. Miseki, *Chem. Soc. Rev.*, 2009, **38**, 253.
17. S. Braslavsky, N. Serpone *et al.*, *Pure Appl. Chem.*, 2011, **83**, 931.
18. K. Maeda and K. Domen, *J. Phys. Chem. Lett.*, 2010, **1**, 2655.

10 Future Applications of Photocatalysis

10.1 Solar Water Splitting

In the early days of the application of photocatalysts, the main research object was to produce clean hydrogen energy by utilizing sunlight to decompose water as stated in Chapter 1. The history of reports on the photocatalysis of water is long, but there were not very many strict experimental reports as stated in Section 9.9. Numerous reports on the photodecomposition of water have been published, and they were reviewed in the literature, for instance in ref. 1. As methods of photocatalytic decomposition of water, particle and electrode systems can be taken up. The former is adequate for obtaining products with larger surface area, and the latter has characteristics that the carrier can be suppressed by applying a bias voltage. It can be also categorized by whether redox can be performed by a single photon excitation, or two photon excitation by which oxidation and reduction are performed in different photocatalytic systems.

10.1.1 One-step Excitation Photocatalysts

It is possible to decompose water to hydrogen and oxygen with TiO_2 under UV light. But the conduction band position of TiO_2 is close to the potential of hydrogen generation. Then, Kudo *et al.* searched the materials which can completely decompose water such as multiple metal oxides possessing d^0 metal ion (*e.g.*, tantalum) as the center element. The conduction band of sodium tantalate ($NaTaO_3$)

Introduction to Photocatalysis: From Basic Science to Applications
By Yoshio Nosaka and Atsuko Nosaka
© Yoshio Nosaka and Atsuko Nosaka, 2016
Published by the Royal Society of Chemistry, www.rsc.org

comprises 5s-orbital of Ta and hydrogen is generated without the aid of a co-catalyst, although NiO is often used as co-catalyst for hydrogen generation to facilitate the activity. Furthermore, by doping La to NiO/NaTaO$_3$ the complete decomposition of water was achieved at high quantum yield of 50%.[2] Because the generated H$_2$ and O$_2$ are reactive, it is best to generate both gases separately. In this reaction system, as observed with TEM, by doping La the produced NaTaO$_3$ becomes fine particulates along with cutting grooves on the surface as shown in Figure 10.1. Because the reduction site can be separated from the oxidation site at the top and the bottom of the grooves, efficient water decomposition became possible.

Water decomposition with visible light is necessary to use solar energy. Domen *et al.* have achieved complete water decomposition by doping ZnO on remarking the nitride of Ga of d^{10} atom. They used the metal nanoparticle of Rh as a catalyst of hydrogen generation, and reported that by covering the surface of Rh with Cr$_2$O$_3$ of 2 nm thickness (see Figure 10.2) the recombination of H$_2$ and O$_2$ could be prevented and, as a result, the activity was enhanced.[3] At the present stage under visible-light irradiation ($\lambda > 400$ nm) with a high-pressure mercury lamp, water is decomposed completely to generate H$_2$ at a rate of about 0.2 mmol h^{-1}. Furthermore, no generation of N$_2$ due to the decomposition of catalysts was observed. However, because only a part of the shorter wavelength range of the visible light region could be utilized, development of photocatalysts using semiconductors such as Ta$_3$N$_5$ whose absorption begins from 600 nm and the band-gap covers the potentials of oxidation and reduction of water is expected.

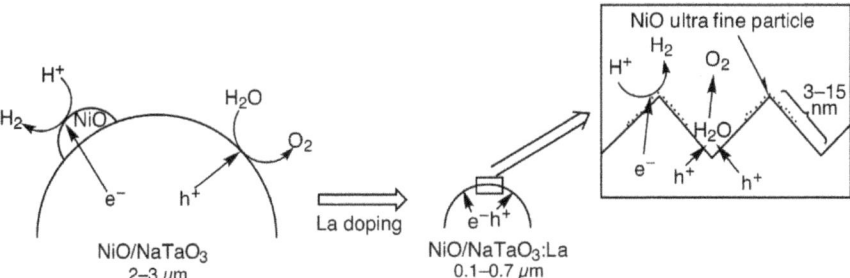

Figure 10.1 Mechanism of highly efficient photocatalytic water splitting over NiO/NaTaO$_3$:La photocatalysts.
Reprinted with permission from T. Kato, K. Asakura and A. Kudo, *J. Am. Chem. Soc.*, 2003, **125**, 3082. Copyright 2003 American Chemical Society.[2]

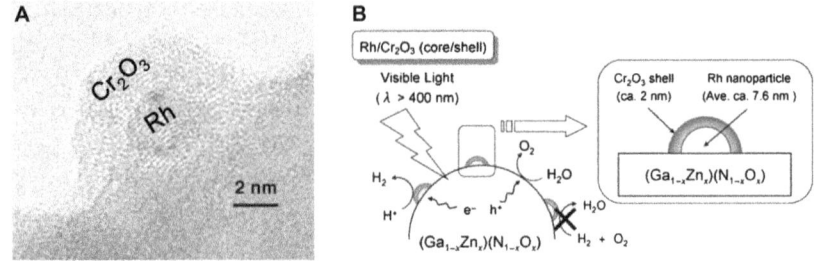

Figure 10.2 (A) HR-TEM images of Rh-loaded $(Ga_{1-x}Zn_x)(N_{1-x}O_x)$ after
photodeposition of the Cr shell. (B) A schematic reaction
mechanism of overall water splitting on Rh/Cr_2O_3-loaded
$(Ga_{1-x}Zn_x)(N_{1-x}O_x)$ and the corresponding processes on sup-
ported Cr_2O_3-Rh nanoparticles.
Reproduced from K. Maeda, K. Teramura, D. Lu, N. Saito,
Y. Inoue and K. Domen, Noble-Metal/Cr_2O_3 Core/Shell Nanoparticles as a Cocatalyst for Photocatalytic Overall Water
Splitting, *Angew. Chem., Int. Ed.*, **45**, 7806–7809 with permission from John Wiley and Sons. Copyright © 2006 WILEY-
VCH Verlag GmbH & Co. KGaA, Weinheim.

10.1.2 Two-step Excitation Photocatalysts (Z-scheme)

For one-step excitation of redox of water stated above, it is difficult to
perform the four-electron reaction efficiently with the energy of visible
light absorption because of the potential loss by the individual surface reactions. Then, the reaction system similar to the Z-scheme of
photosynthesis in which the oxidation and the reduction proceed in
the separate reaction system as shown in Figure 10.3 is attempted.[4] In
this system, redox couples of Fe^{3+}/Fe^{2+}, Co^{3+}/Co^{2+}, or IO_3^-/I^- are
used as mediators to connect the oxidative and the reductive systems.
For high efficiency, it is necessary to manage a device to separate
them with porous membranes so that the reduced mediator is not
oxidized in the oxidative system or that the oxidized mediator is not
reduced in the reductive system.

A number of combinations of semiconductor photocatalysts for the
oxidative and reductive systems are proposed,[1] where the metal oxide
is not decomposed but stable. An example is shown in Figure 10.4.
For the oxidative system $BiVO_4$ semiconductor powder is used, while
for the reductive system $SrTiO_3$ powder is used by depositing Ru and
doping Rh ($Ru/SrTiO_3$:Rh). As shown in Figure 10.4, by using 120 mL
of 0.5 mM $[Co(bpy)_3]SO_4$ at pH 3.8 as a mediator, H_2 and O_2 can be
generated separately by irradiation with a Xe lamp with a glass filter
cutting the wavelength shorter than 420 nm. The photocatalyst is

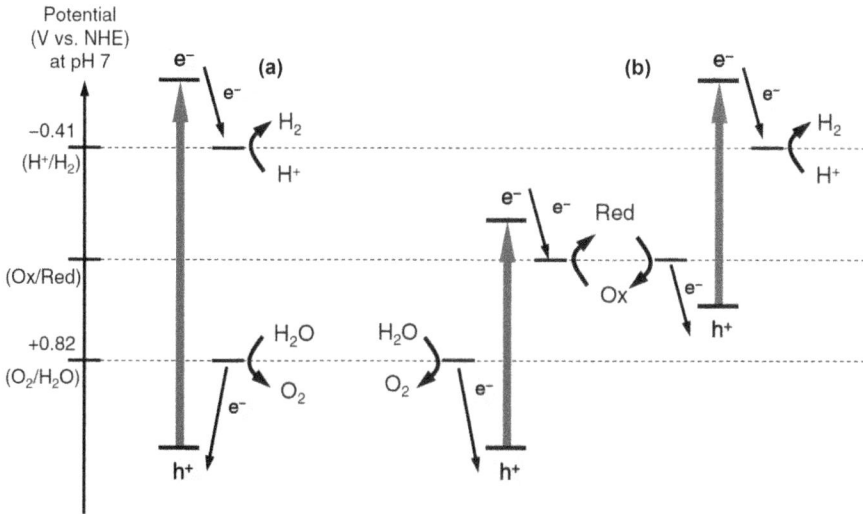

Figure 10.3 Forward and backward reactions in the two-step photoexcitation system.
Reprinted with permission from ref. 4 (*Bull. Chem. Soc. Jpn.*). Copyright 2011 the Chemical Society of Japan.

placed in a separate vessel and the mediator can only move using a membrane filter as shown in Figure 10.4B.

Higher activity can be achieved by mixing the particles than by separating $Ru/SrTiO_3:Rh$ powders from $BiVO_4$.[5] Owing to the contact of two particles, Rh doped in $SrTiO_3$ is considered to become an effective acceptor of the excited electrons of $BiVO_4$. Namely, like a combined photocatalyst shown in Figure 7.14A in Chapter 7, a CB electron at the oxidative side and a VB hole at the reductive side are considered to combine efficiently.

It is known that the Z-scheme is formed by using the nanoparticle deposited with Ir metal or RGO (Reduced Graphene Oxide) instead of a mediator.[6,7] By forming the Z-scheme, the semiconductors whose bandgap is exclusively too small to decompose water can be utilized, which opens the possibility of the further practical applications. In addition, with this scheme, the system can also be formed by taking account of dye-sensitization. The disadvantage of this method is that the system for which the rate of photoabsorption is lower becomes rate determining. Therefore, without the balance of excitation, undesired reactions can easily take place in the excessively excited system. Furthermore, the efficiency of light utilization becomes not more than 50%.

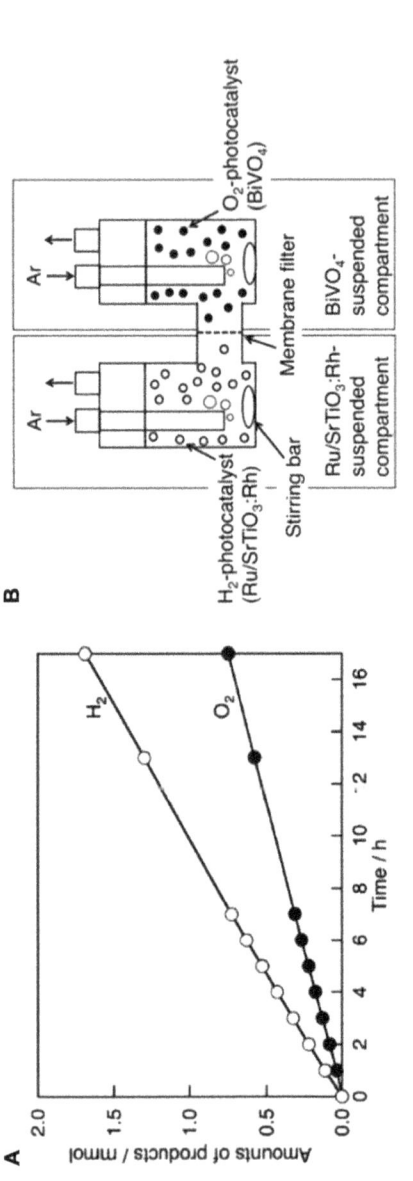

Figure 10.4 (A) Overall water splitting under visible light irradiation by (A) mixed systems of Ru/SrTiO$_3$:Rh and BiVO$_4$ powder under visible light irradiation. (B) Two-compartment-reactor for H$_2$ evolution separated from O$_2$ evolution.
Reprinted with permission from Y. Sasakai, H. Kato and A. Kudo, *J. Am. Chem. Soc.*, 2013, **135**, 5441. Copyright 2013 American Chemical Society.[5]

10.1.3 Photoelectrodes

The principle of water decomposition with semiconductor electrodes has become known as the Honda–Fujishima Effect (Section 4.3.1). Because TiO_2 absorbs only a few % of solar energy, methods for TiO_2 to absorb visible light, or utilization of the semiconductors whose bandgap is narrow, are under consideration. Because the redox potentials of the semiconductors with narrow bandgap become low, it is necessary to boost the potential by a photovoltaic (PV) cell as shown in Figure 10.5A. The photoanode here is an n-type semiconductor with a VB edge relatively positive to the water oxidation potential. The PV cell of the photogenerated electrons allows water reduction at the cathode surface.

As shown in Figure 10.5B, it is effective to allow the redox reactions by two photons at n- and p-type semiconductors separately. Because for the semiconductors with relatively wide bandgap the light of the longer wavelength is not absorbed but transmits, absorbing the transmitted light with a semiconductor electrode of narrow bandgap is an effective method to use.[8] An example will be shown for the CO_2 reductive electrode system in Section 10.7. The electrode catalysts which are highly active for the reduction and oxidation of water have been known for a long time.[9] For semiconductor photoelectrodes, each surface is often deposited with these catalysts.

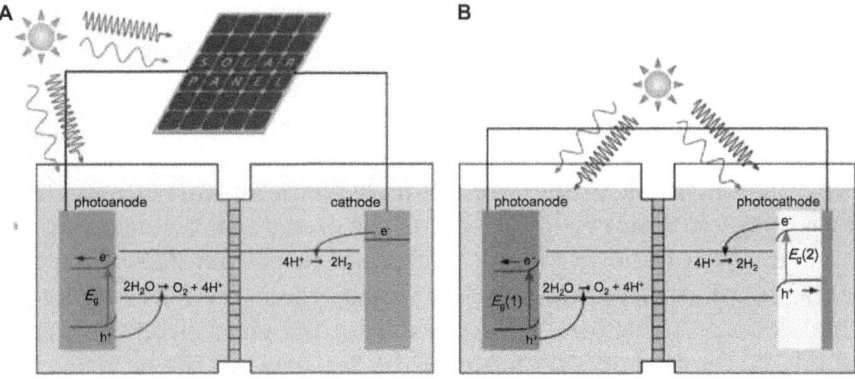

Figure 10.5 Systems for photoelectrochemical solar water splitting. (A) A single-photoelectrode PEC cell assisted by a photovoltaic (PV) cell. (B) A two-photoelectrode PEC cell in which the photoanode and photocathode are suitably matched n-type and p-type semiconductors, respectively.
Reproduced from ref. 8 (*Energy Environ. Sci.*, 2010).

The dependency of photocurrent on wavelength intrinsically corresponds to the absorption spectrum. It is more practical to evaluate the photoelectrodes by the current efficiency against the amount of incident photons than against the amount of absorbed photons, which is often called incident photon-to-current efficiency (IPCE). IPCE is the ratio of the number of electrons of photocurrent $i(\lambda)$ (mA) to the number of photons irradiated at any wavelength λ, as expressed by eqn (10.1).

$$\text{IPCE}(\lambda) = \frac{i(\lambda) \times 1240}{I(\lambda) \times \lambda} \times 100(\%) \tag{10.1}$$

where $I(\lambda)$ (mW) represents the incident light intensity.

Several photoanodes and photocathodes whose absorption spectrum sufficiently spans the solar spectrum with large IPCE are known. For instance, the photocathode, where p-Cu_2O is covered with Al doped ZnO film 21 nm thick, which is further covered with TiO_2 11 nm thick, and deposited with Pt particle as a catalyst, is reported.[10] When the electrode potential is set to 0 V (*vs.* RHE), the photoreductive current of 7.8 mA cm^{-2} can be obtained under the simulated solar light of AM1.5 and Faradic efficiency of H_2 generation, which is the ratio of the current used for the synthesis of H_2 to the total current, and is close to 100%. The problem is the lifetime of the electrode.[10] The photoanodes which are fabricated to coat Si, GaAs, and GaP semiconductors with TiO_2 to be stabilized are also known. For instance, for n-GaP electrode coated with Ni 2 nm thick and TiO_2 118 nm thick, IPCE is more than 60% under the wide range of the visible light wavelength and the photoreductive current of 15 mA cm^{-2} is obtained under the simulated solar light of AM1.5.[11]

Because these electrodes are fabricated with an ALD (atomic layer deposition) technique, it is difficult to prepare photoelectrodes of large area. On the other hand, photocatalyst powder can be used to fabricate photoelectrodes. For example, after the electrophoretic deposition of TaON (or Ta_3N_5) powder on the glass electrode coated with FTO (fluorine doped tin oxide), the gap between the particles and the FTO substrate was filled with $TaCl_3$ solution and necked with heat treatment under NH_3 flow. For these electrodes IPCE of 31% was achieved under light irradiation at 500 nm.[12] Thus, in the future, the fabrication of the semiconductor photoelectrode of large area is expected.

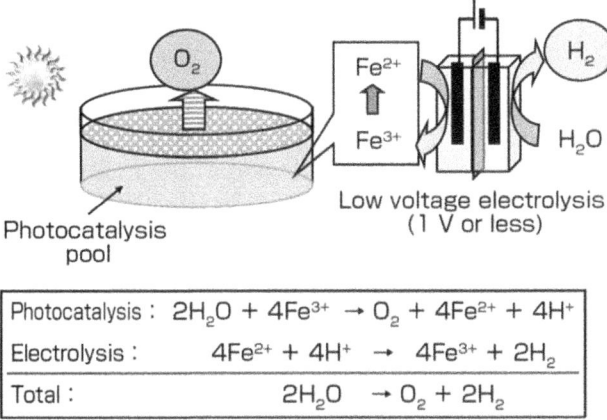

Photocatalysis :	$2H_2O + 4Fe^{3+} \rightarrow O_2 + 4Fe^{2+} + 4H^+$
Electrolysis :	$4Fe^{2+} + 4H^+ \rightarrow 4Fe^{3+} + 2H_2$
Total :	$2H_2O \rightarrow O_2 + 2H_2$

Figure 10.6 The photocatalysis pool made of a plastic bag which consists of a layer of photocatalyst powder and electrolytic aqueous solution containing the redox mediator of Fe ions. Reproduced from ref. 13. Courtesy of AIST.

10.1.4 Photocatalysis–Electrolysis Hybrid Systems

The combination of a powder photocatalytic system and an electrode system is also contrived, as illustrated in Figure 10.6. In this hybrid system, the powder photocatalyst with Fe^{3+}/Fe^{2+} as a mediator oxidizes water and simultaneously reduces Fe^{3+} to Fe^{2+}, which reduces water with the electrode under a bias voltage of 0.8 V. In this case the total cost for H_2 gas generation is provisionally calculated (based on the Japanese market) to be 0.18 Euro m^{-3} when the quantum yield of the photocatalyst whose cost is twice of that of WO_3 is 3%, using the catalyst pool of 3 km^2 area made of polyethylene under mean daylight hours of 4 h, and taking into account the estimated depreciation of the instrument of 10 years. For simple electrolysis of water the price is 0.30 Euro m^{-3}, while the realistic price is 0.22 Euro m^{-3}.[13] Thus, with this hybrid system using photocatalyst for H_2 production the cost becomes 17% less than the realistic price.

10.2 Carbon Dioxide (CO₂) Reduction

Photocatalytic reduction of CO_2 is gathering much attention due to global warming owing to the increase of atmospheric CO_2 and a lot of relevant work has been published.[14,15] Many reports dealt with

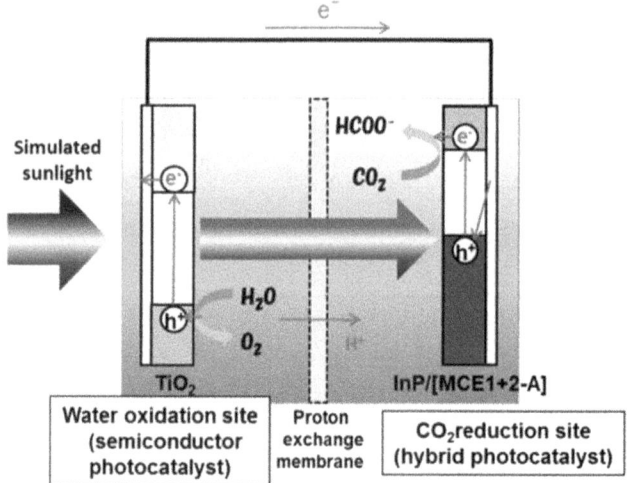

Figure 10.7 Schematic illustration of the tandem-cell reactor for reduction of CO_2.
Reprinted with permission from S. Sato, T. Arai and T. Morikawa, *Inorg. Chem.*, 2015, **54**, 5105. Copyright 2015 American Chemical Society.[17]

catalysts using a sacrificial donor. As stated in Chapter 1 (Section 1.6), the reports on CO_2 reduction have been published for a long time, but the number of reports relevant to the reductive products were quite small. In addition the discussion on oxidative reactions was often omitted. For practical use the generation of oxygen on the oxidation of water is a prerequisite.

The reduction of CO_2 is much more difficult than that of water. This is understandable from the facts that the reductive potential of a proton is 0.0 V, while the potential of one electron reduction for CO_2 is −1.9 V *vs.* SHE (Figure 4.4). Then, it is necessary to develop reductive catalysts with which CO_2 is reduced in water without reducing water. For electrode systems, the p-type NiO semiconductor electrode deposited with Ru(II)–Ru(I)–multinuclear metal complex[16] and the electrode in which Ru complex is deposited on p-type InP as shown in Figure 10.7 [17] were reported.

On the other hand, for a powder system, as stated in the section on water splitting (Section 9.9), if it were not shown experimentally what is oxidized along with the reduction of CO_2, the practical use for CO_2 reduction would be far from the practical application. One of the few reports that showed this experimentally would be on the system with Ag-deposited $BaLa_4Ti_4O_{15}$.[18] As shown in Figure 10.8A, because

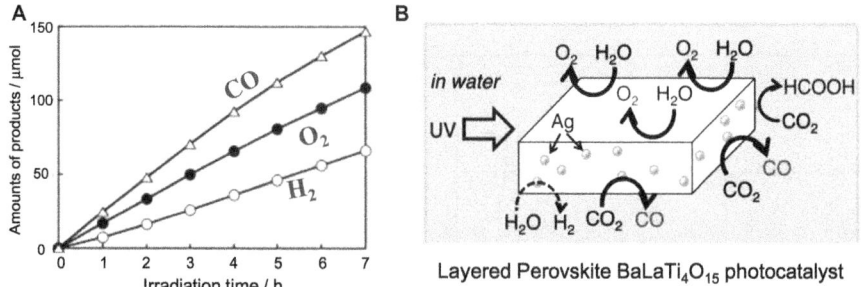

Figure 10.8 Liquid-phase CO_2 reduction over $BaLa_4Ti_4O_{15}$ photocatalyst with Ag cocatalyst in CO_2 flow system irradiated with a 400 W high-pressure mercury lamp. (A) The amounts of products with irradiation time. (B) Reaction mechanism. Reprinted with permission from K. Iizauka and A. Kudo *et al.*, *J. Am. Chem. Soc.*, 2011, **133**, 20863. Copyright 2011 American Chemical Society.[18]

the amount of O_2 generation is half of those of $CO_2 + H_2$, the generation of O_2 corresponding to the reductive reaction can be confirmed.

10.3 Sensor Applications

The importance of sensors is growing to acquire precise information on various categories because social situations largely depend on the rapid exchange of a variety of information. Properties such as high sensitivity, fast response, and good selectivity are prerequisites for sensors.[19] The representative researches on sensor applications would be those on an oxygen sensor and various biosensors.

10.3.1 Chemical Oxygen Demand (COD) Sensor

With a thin cell-reactor constructed by using a TiO_2 nanotube array fabricated by electrochemical anodization as shown in Figure 10.9A, COD of waste water can be measured with a small quantity of the sample. Although at high concentration the light emission efficiency becomes different depending on the kinds of organic compound, when the concentration is less than several hundred mg L^{-1}, the linear relationship between the photocurrent and COD of the solution holds independent from the kinds of organic compound dissolved.[20]

A probe type COD sensor in which the light source is unified is shown in Figure 10.9B, where the TiO_2 working electrode is 1 cm from the UV-LED lamp across the sample solution, and the counter and the

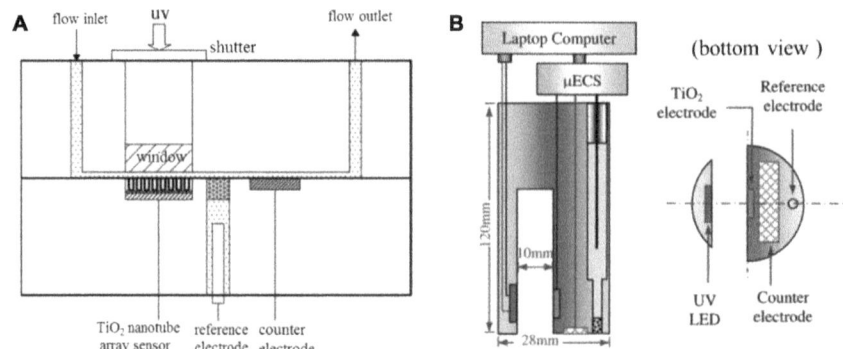

Figure 10.9 (A) Photoelectrocatalytic thin-cell COD sensor. (B) Schematic diagram of the photoelectrochemical probe for rapid COD determination. Reproduced (A) from Q. Zheng, B. Zhou, J. Bai, L. Li, Z. Jin, J. Zhang, J. Li, Y. Liu, W. Cai and X. Zhu, Self-organized TiO_2 nanotube array sensor for the determination of chemical oxygen demand, *Adv. Mater.*, **20**, 1044–1049. Copyright (2008), with permission from John Wiley and Sons. © 2008 WILEY-VCH Verlag GmbH & Co. KGaA, Weinheim, and (B) with permission from S. Zhang, L. Li and H. Zhao, *Environ. Sci. Technol.*, 2009, **43**, 7810. Copyright 2009 American Chemical Society.[21]

reference electrodes are held in the probe. On inserting this probe in the solution, the photocatalytic decomposition of organic compound takes place and the oxidation current is measured.[21]

10.3.2 Biosensors

Various photoelectrochemical biosensors are known.[22] For instance, the sensitive detection of glutathione using the IrO_2–hemin–TiO_2 nanowire arrays has been developed. Single-crystalline TiO_2 nanowires are synthesized by a hydrothermal reaction, followed by the surface functionalization of about 3 nm thick hemin and about 1–2 nm diameter IrO_2 nanoparticles. The sensitivity achieved is about 10 nM in buffer, which is comparable to or even better than that obtained with most of the existing glutathione detection methods. The high sensitivity was brought about by the effect of co-catalysis of IrO_2 clusters on the surface and by the efficient photoinduced charge separation by hemin–TiO_2 composite semiconductor.

The sensing of DNA conducted by detecting the chemiluminescence with a localized photoelectrochemical cell is also proposed.[23] As shown in Figure 10.10, chemiluminescence takes place by H_2O_2,

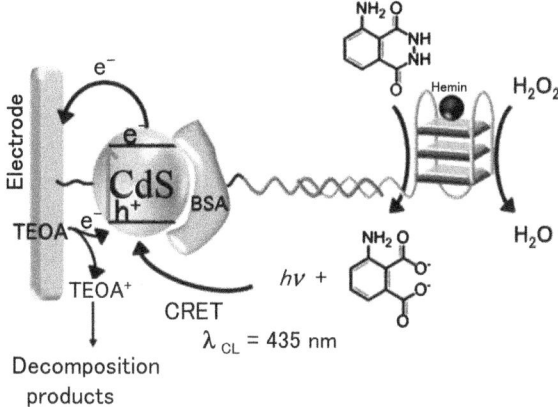

Figure 10.10 Analysis of a target DNA by a sandwich-type nucleic acid assay on the CdS-BSA conjugate associated with an electrode using the hemin/G-quadruplex as a catalytic label for the photocurrents generated by Chemiluminescence Resonance Energy Transfer (CRET).
Reprinted with permission from E. Golub and I. Willner *et al.*, *J. Phys. Chem. C*, 2012, **116**, 13827. Copyright 2012 American Chemical Society.[23]

then CdS is excited to flow electric current. Where CdS quantum dot deposited on the electrode substrate adsorbs DNA-functionalized BSA (Bovine Serum Albumin). The DNA can catch the hemin modified DNA to oxidize luminol. By adsorbing DNA-functionalized Glucose Oxidase (GOx) on CdS, glucose and oxygen can be used in place of H_2O_2. Triethanolamine (TEOA) works as a sacrificial electron donor for the oxidation current of CdS. For this DNA sensor, CdS photocatalyst is utilized as a photodetector.[23]

10.4 Biomedical Applications

Despite the intensive studies of TiO_2 in various fields, the involvement of TiO_2 in biomedical applications is relatively new. The progress of various biological and biomedical applications of TiO_2 would promise future medical innovations in molecular medicine.[24] For therapeutic applications of photocatalysts, the major problem is that the light should be applied in tissues. Figure 10.11 shows absorption of chromophores present in human tissues.[25] Light below 650 nm in wavelength is too strongly absorbed by mainly hemoglobin and above 950 nm too strongly by water. Therefore, the optical window for effective irradiation is about 650–950 nm.

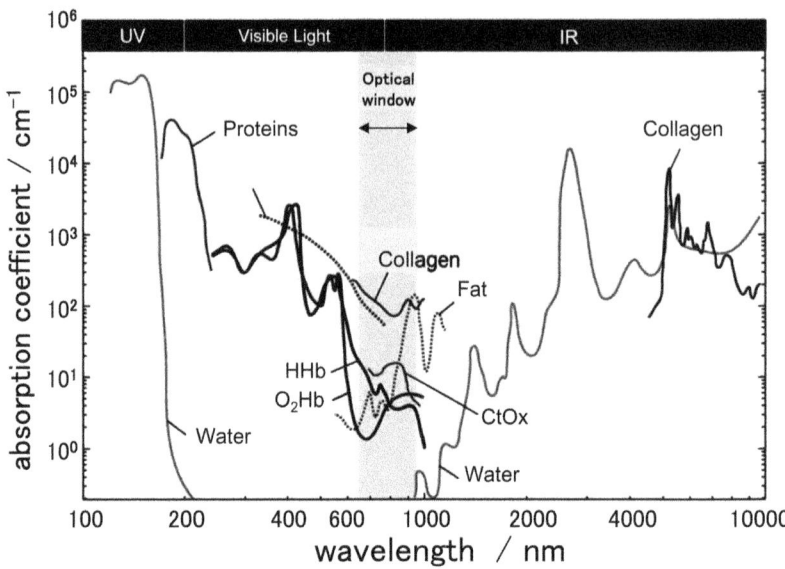

Figure 10.11 Absorption for different chromophores present in human tissue. Shown are the spectra for oxy and deoxy hemoglobins (O_2Hb and HHb), proteins, water, collagen, fat, and cytochrome oxidase (CtOx) in the concentration of mM.
Reprinted from F. Scholkmann, S. Kleiser, A. J. Metz, R. Zimmermann, J. M. Pavia, U. Wolf and M. Wolf, A review on continuous wave functional near-infrared spectroscopy and imaging instrumentation and methodology, *NeuroImage*, **85**, 6–27. Copyright (2014), with permission from Elsevier.

Enediols such as dopamine are coordinated to TiO_2 and present a large absorption in the visible wavelength region accompanied with the electron transfer to the conduction band. Thus, biological applications for enediols attached with various functional groups are proposed.[24] In particular, the high surface-to-bulk ratio of metal oxide nanomaterials offers a unique possibility of linking multiple biological molecules of probes to the surface of each nanoparticle employing precise engineering of their surface. Nanoparticles modified with "conductive leads" and integrated with antibodies established effective light-induced crosstalk across the interface between biomolecules and metal oxide nanoparticles. Furthermore, functionally integrated TiO_2–antibody complexes retain the photocatalytic properties of nanoparticles and recognition properties of monoclonal antibodies. The ability of TiO_2 nanoparticles to initiate electron-transfer reactions within DNA molecules might promise the additional biomedical revolution occurring in the stem cell arena.[24]

10.5 Organic Synthesis

To avoid the further addition of harmful chemicals into the environment, new "green" routes to produce chemicals are expected in chemical industries. Heterogeneous photocatalysts can play an important role in developing new technologies utilized at room temperature under atmospheric pressure. For example, the potential and applications of solid photocatalysts for the selective transformation of biomass-derived substrates are discussed.[26]

The preparatively valuable photoreactions are easily conducted and the heterogeneous photocatalyst can be conveniently separated from the products. They are promising examples of "green chemistry", because they do not produce waste materials and can be driven by solar light. In semiconductor photocatalysis, the resulting primary redox products are radicals, which undergo regio- and stereo-selective C–C and C–N coupling to the final products as was shown for CdS photocatalysts.[27] The recovery of the products by coupling photocatalysis with pervaporation membrane technology seems to be a solution of the problems for future industrial applications. As far as photocatalysis is concerned, not many papers aiming at the production of chemicals have been published. This technology looks convenient only for a suitable choice of the electrode catalysts. Actually, the occurrence of the reduction and oxidation photoreactions in separated sides is positive, but the difficulty of working in aqueous environments is of course an issue.[28]

10.6 Application of Superhydrophilicity

The superhydrophilicity of TiO_2 has been practically utilized for door mirrors of cars to prevent the formation of water droplets.[29] Such a property could also be used for printing plates and cooling of buildings.

10.6.1 Printing Plates

TiO_2 photocatalyst-based superhydrophobic–superhydrophilic patterns are fabricated *via* an ink-jet technique and offset printing. As shown in Figure 10.12 the process consists of five key steps: (1) TiO_2 coating of the plate, (2) surface modification with self-assembled monolayers (SAMs), (3) formation of aqueous UV light-resistant ink patterns by an ink-jet technique, (4) photocatalytic decomposition of

Top   Cross section

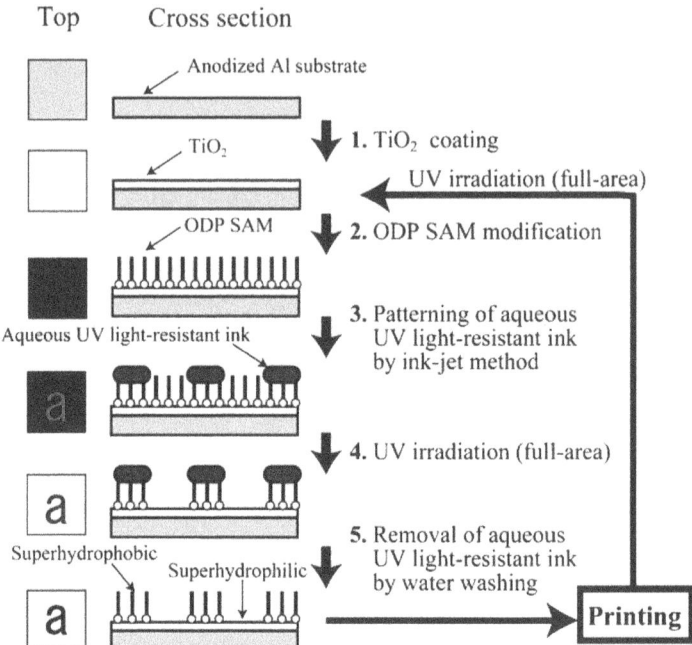

Figure 10.12   Schematic diagram of TiO₂-based superhydrophobic–superhydrophilic patterning by an ink-jet technique and reusing processes.
Reprinted from S. Nishimoto, A. Kubo, K. Nohara, X. Zhang, N. Taneichi, T. Okui, Z. Liu, K. Nakata, H. Sakai, T. Murakami, M. Abe, T. Komine and A. Fujishima, TiO₂-based superhydrophobic–superhydrophilic patterns: Fabrication *via* an ink-jet technique and application in offset printing, *Appl. Surf. Sci.*, **255**, 6221–6225. Copyright (2009), with permission from Elsevier.

SAMs and the surface conversion to the superhydrophilic state, and (5) removal of the aqueous ink patterns by water washing. It is particularly noteworthy that the wettability pattern can be quickly formed on the plate without the use of a photomask.

The reusable TiO₂-based printing plate showed sufficient durability against web pressing and clear color printing is possible with a resolution of 133 LPI (lines per inch). Little organic waste fluid is discharged in the fabrication process because patterning is possible in a dry process without the use of a photoresist, while in the conventional printing process the organic effluent is discharged. Consequently, the TiO₂-based printing plate is an environmentally friendly material.[30]

10.6.2 Cooling Systems of Buildings

The reduction of cooling energy consumption for rooms is attempted by making a thin film of water on the surface of the building to utilize the latent heat of evaporation.[31] On sprinkling a small amount of water over a glass plate standing upright, when the water contact angle of the surface becomes less than 10°, the wet area of the surface drastically increases. When the contact angle becomes 0°, more than 90% of the surface can be covered with the water film 0.1 mm thick.

The surface temperature of the glass on sprinkling (300 mL min^{-1}) was measured on a demonstration test. Glass curtain walls 11 m wide and 5 m high whose surface is coated with TiO$_2$ are placed at the west and north sides of the building facing the road. As shown in Figure 10.13A, by flowing water from 9 a.m. to 6 p.m., the temperature of the glass was found to decrease by 14 °C at maximum and even on average by 8 °C.

Figure 10.13B shows the plot of the gas consumption of the air conditioner working to control the room temperature to be 25 °C as a function of the outdoor temperature. The energy consumption is evidently more suppressed for higher outdoor temperature by sprinkling. It is implied that on average 12% of energy can be reduced by sprinkling when the outdoor temperature is 30 °C.[31]

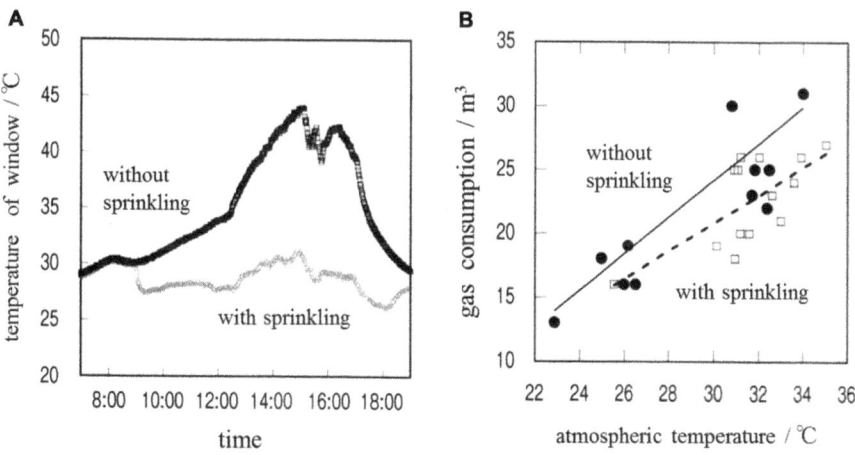

Figure 10.13 Effect of sprinkling on a TiO$_2$-coated glass wall. (A) Temperature change of glass wall and (B) gas consumption for air conditioning to 25 °C in the room.
Reproduced from ref. 31 (*Zukai Hikarishokubai no Subete*, 2012) with permission from Ohmsha Ltd.

10.7 Industrial Applications

10.7.1 Oxygen Indicator for Food Packaging

The detection of oxygen is necessary in various food industries such as an airtight container for the maintenance of freshness. As for colorimetric oxygen indicators, although some commercial systems are available, they suffer from high retail costs and a lack of reliability. Methylene blue (MB) is known to become colorless leuco MB by photocatalytic reduction, but it returns to blue-colored MB with O_2. This reaction is utilized for the O_2 indicator as an O_2 smart plastic film. Low-density polyethylene (LDPE) is used to prepare the film with nanoparticulate pigment particles that consist of TiO_2 coated with MB and a sacrificial electron donor, threitol. The blue-colored indicator is photobleached within 90 s at 4 mW cm^{-2} photoirradiation and recovers in 2.5 days in air under ambient conditions. The key parameter to detect O_2 is the increase in b* factor which is recommended by the CIELAB color scale as a standard of blue. The O_2 smart plastic film is waterproof and reusable, thus it is incorporated into food packaging.[32]

10.7.2 Decontamination of Soil Pollutants

Soil contamination by volatile chlorinated organic compounds (VCOC) such as trichloroethylene which was heavily used as a volatile solvent is a problem. In Figure 10.14 the schematic diagram of VCOC

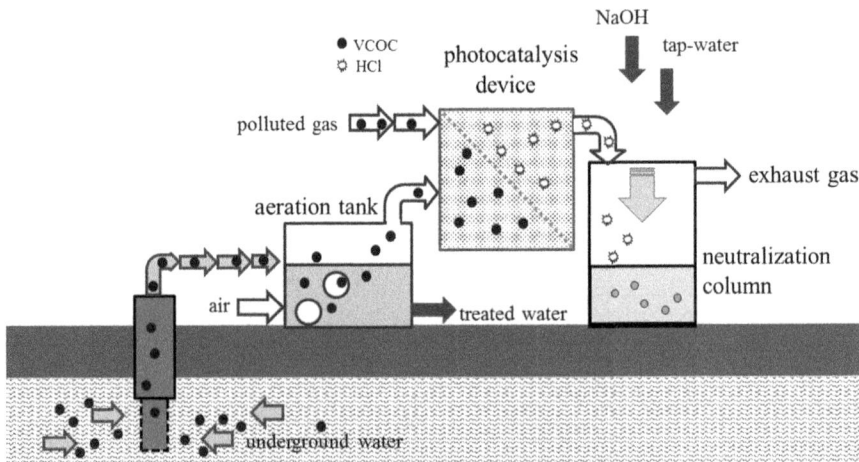

Figure 10.14   Decontamination system for soil polluted with volatile chlorinated organic compound (VCOC).
Reproduced from ref. 33 with permission of Adeka Engineering & Construction Corporation.

treatment by photocatalytic techniques is shown. In the unsaturation zone above the ground-water level, the pollutant substances are collected into the vapor phase by soil vapor extraction, and in the saturation zone under the ground-water level, the pollutant substances are collected into the vapor phase by an aeration method, then they are processed with photocatalytic treatments.

Previously, activated carbon (charcoal) was used instead of photocatalyst treatments in the decontamination system. When cleaning the substances diffused in the soil it takes long time, and it requires effort and costs for maintenance such as the replacement and disposal of the charcoal. On the other hand, by using an apparatus for photocatalysis which decomposes and disposes of the substances, the maintenance problems can be substantially reduced. In Japan, 70 apparatuses have already been constructed in 2015.[33]

10.7.3 Treatment of Hydroponics Waste

To dispose of the agricultural effluent in the hydroponics (solution culture) facility of tomatoes, the photocatalysis system was introduced and the test culture has begun. For the hydroponics facility of tomatoes, a waste water treatment tank whose bottom is covered with the TiO_2-coated ceramic porous plate is placed outside the greenhouse.

As shown in Figure 10.15, firstly the nourishing solution is supplied from the liquid supply tank to the palm husk culturing

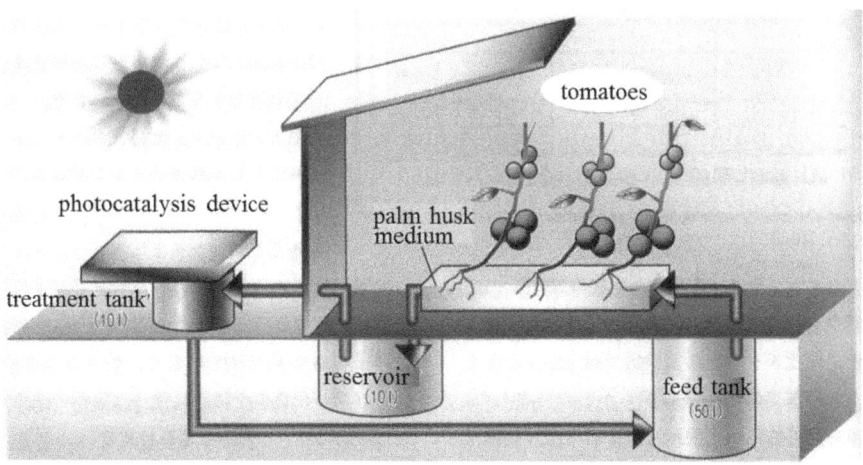

Figure 10.15 Photocatalytic waste treatment in hydroponics of tomatoes. Reproduced from ref. 34 (*Zukai Hikarishokubai no Subete*, 2012) with permission from Ohmsha Ltd.

medium where tomatoes are planted.[34] Then, the excessive nourishing solution which is not absorbed by tomatoes is collected in the storage tank as sewerage. When the sewerage exceeds a certain amount, it is transferred to the process tank located outside the greenhouse and photocatalytic treatments are processed. The processed solution is restored in the storage tank and reused. With this circulation system the hydroponics of tomatoes were actually conducted and, after 10 months observation, the TOC in the storage tank was found to be 7–28 mg L^{-1}, while it was 21–777 mg L^{-1} when the waste fluid was not processed with photocatalyst. Thus, it was confirmed that plants and organic compounds eluted from the organic culture are efficiently decomposed under sunlight. Furthermore, with this processing system using a photocatalyst both the growth and yield exceeded those cultivated without a photocatalytic process.[34]

10.7.4 Treatment of Residual Pesticides

To prevent the disease of agricultural plants by noxious insects, usually pesticides are used. As for the residual pesticides on the plants after usage, it is desirable to rapidly decompose them in order to remove them. The pesticides mixed with the same amount of TiO$_2$ were sprayed and the residual quantity of the pesticides was measured with time. Then, 7 days later the residues were not detected and became less than that of the safety standard value for residual pesticide concentration. On the other hand, without TiO$_2$ it was confirmed that a small amount of the residue was detected even 14 days later.[35]

The seed disinfection of paddy-field rice aiming at pest control is indispensable for stable production, but it is laborious to process agricultural effluent after disinfection. Namely, such processes are taken that the agricultural waste in the effluent is adsorbed on the active charcoal and filtered after the addition of coagulant, then the residual is dried, which is processed by manufacturers. Hence, a simple treatment process of the agricultural effluent with the aid of photocatalyst and sunlight is proposed.

TiO$_2$-coated porous ceramic is spread in the tank (1.5 m×1.5 m area and 5 cm high), which is covered with a UV-transmitting film to prevent evaporation. As for a bactericide of ipconazole, TOC is 2.5 mg L^{-1} 10 days later without photocatalysts, but with photocatalysts it becomes under the detection limit (0.01 mg L^{-1}). The cost for the TiO$_2$ used is calculated to be about 3% of that for pesticide.

The practical application is expected as a system for safe seed disinfection.[36]

10.7.5 Disinfection of Sea Water

Sea water is used in fish markets and fishing harbors because, when tap water is used to wash and preserve fish, the fish color changes and the freshness is lost. For safety and health, it is necessary to remove saprophytes in sea water, but for disinfection by chemical agents, there is a problem with the safety of residuals. Therefore, disinfection is often performed by UV light. However, in spite of disinfection by UV light, the photorecovery phenomenon that bacteria again proliferate under the illumination of visible light takes place.

To process a large quantity of sea water in a short time and to supply the disinfected sea water without photorecovery, a system in which photocatalytic treatment combined with the UV light and hydrogen peroxide treatments is contrived. In this system, filtrated raw sea water is first treated with UV sterilization for the use of floor cleaning and chilled storage. Then, after treatment with hydrogen peroxide the treated sea water successively flows to a photocatalysis reactor. The reactor cell is filled with alumina filter coated with TiO_2 and UV light is irradiated on the glass window of the cell. It is reported that safe disinfected sea water without photorecovery can be supplied 1–3 m^3 day^{-1}, under consuming hydrogen peroxide of 6–14 mg L^{-1} for 10 h.[37]

References

1. X. Li, J. Yu *et al.*, *J. Mater. Chem. A*, 2015, **3**, 2485.
2. T. Kato, K. Asakura and A. Kudo, *J. Am. Chem. Soc.*, 2003, **125**, 3082.
3. K. Maeda, K. Domen *et al.*, *Angew. Chem., Int. Ed.*, 2006, **45**, 7806.
4. R. Abe, *Bull. Chem. Soc. Jpn.*, 2011, **84**, 1000.
5. Y. Sasakai, H. Kato and A. Kudo, *J. Am. Chem. Soc.*, 2013, **135**, 5441.
6. Q. Wang, K. Domen *et al.*, *Chem. Mater.*, 2014, **26**, 4144.
7. A. Iwase, R. Amal *et al.*, *J. Am. Chem. Soc.*, 2011, **133**, 11054.
8. J. Sun, D. K. Zhong and D. R. Gamelin, *Energy Environ. Sci.*, 2010, **3**, 1252.
9. Y. Jiao, S. Z. Qiao *et al.*, *Chem. Soc. Rev.*, 2015, **44**, 2060.
10. A. Paracchino, E. Thimsen *et al.*, *Nat. Mater.*, 2011, **10**, 456.

11. S. Hu, N. S. Lewis *et al.*, *Science*, 2014, **344**, 1005.
12. M. Higashi, K. Domen and R. Abe, *Energy Environ. Sci.*, 2011, **4**, 4138.
13. K. Sayama and Y. Miseki, *Synthesiology Eng. Ed.*, 2014, **7**, 79.
14. S. Das and W. M. A. W. Daud, *RSC Adv.*, 2014, **4**, 20856.
15. L. Yuan and Y. Xu, *Appl. Surf. Sci.*, 2015, **342**, 154.
16. G. Sahara, O. Ishitani *et al.*, *Chem. Commun.*, 2015, **51**, 10722.
17. S. Sato, T. Arai and T. Morikawa, *Inorg. Chem.*, 2015, **54**, 5105.
18. K. Iizauka, A. Kudo *et al.*, *J. Am. Chem. Soc.*, 2011, **133**, 20863.
19. J. Bai and B. Zhou, *Chem. Rev.*, 2014, **114**, 10131.
20. Q. Zheng and B. Zhou, *Adv. Mater.*, 2008, **20**, 1044.
21. S. Zhang, L. Li and H. Zhao, *Environ. Sci. Technol.*, 2009, **43**, 7810.
22. A. Devadoss, C. Terashima *et al.*, *J. Photochem. Photobiol., C*, 2015, **24**, 43.
23. E. Golub, I. Willner *et al.*, *J. Phys. Chem. C*, 2012, **116**, 13827.
24. T. Rajh *et al.*, *Chem. Rev.*, 2014, **114**, 10177.
25. F. Scholkmann, M. Wolf *et al.*, *NeuroImage*, 2013, **85**, 6.
26. J. C. Colmenares and R. Luque, *Chem. Soc. Rev.*, 2014, **43**, 765.
27. H. Kisch, *Angew. Chem., Int. Ed.*, 2013, **52**, 812.
28. V. Augugliaro, G. Palmisano *et al.*, *J. Phys. Chem. Lett.*, 2015, **6**, 1968.
29. S. Banerjee, D. D. Dionysiou and S. C. Pillai, *Appl. Catal., B*, 2015, **176**, 396.
30. S. Nishimoto *et al.*, *Appl. Surf. Sci.*, 2009, **255**, 6221.
31. K. Sunada *et al.*, *Zukai Hikarishokubai no Subete (Illustrated All of the Photocatalysis)*, ed. K. Hashimoto and A. Fujishima, Ohmusha Inc., Tokyo, 2012, p. 175.
32. A. Mills *et al.*, *Analyst*, 2012, **137**, 106.
33. Home page of Adeka Co. http://www.adkeng.co.jp/product/ecoclean_opt.html.
34. Y. Miyama, *Zukai Hikarishokubai no Subete (Illustrated All of the Photocatalysis)*, ed. K. Hashimoto and A. Fujishima, Ohmusha Inc., Tokyo, 2012, p. 238.
35. K. Kusano, *Zukai Hikarishokubai no Subete (Illustrated All of the Photocatalysis)*, ed. K. Hashimoto and A. Fujishima, Ohmusha Inc., Tokyo, 2012, p. 232.
36. Y. Miyama, *Zukai Hikarishokubai no Subete (Illustrated All of the Photocatalysis)*, ed. K. Hashimoto and A. Fujishima, Ohmusha Inc., Tokyo, 2012, p. 235.
37. H. Noguchi *et al.*, *Photocatalysis*, 2003, **9**, 46.

Subject Index